THEY STARED AT THE SUN

THEY STARED AT THE SUN
CONTEMPORARY DEVELOPMENTS IN EVOLUTIONARY THEORY

DAVID SPOONER

To order additional copies of this book, contact:
Xlibris LLC
0-800-056-3182
www.xlibrispublishing.co.uk
Orders@xlibrispublishing.co.uk
609232

CONTENTS

DEDICATIONS

for Hyatt Carter, fount of international spiritual insights

in memory of Norman O. Brown, the great source of the associative method

with thanks to Marion O'Neil, Scotland's skilled contemporary archeological illustrator, without whose decades-long personal support, this book would never have been written.

FOREWORD

This book has been written in the belief that the scientists have failed to take into account those areas of human activity and original achievement that actually define the species. The worlds of literature and music both offer "messages" that free the human species from its physical limits.

Alfred Russel Wallace had insisted against Darwin that "the existence in man of something which has not derived from his animal progenitors . . . a spiritual essence or nature, capable of progressive development under favorable conditions."[1] As I wrote in my *The Insect-Populated Mind: how insects have influenced the evolution of consciousness* (2005), Wallace's concentration on insects, together with his insistence that the nature of the human brain lay in "the *evasion* of specialization"[2] suggest a more profound gulf than either Wallace or Darwin found it tactically opportune, in the light of the newness of their central theory and the opposition aroused, to admit between their views.

The full instatement of Wallace as co-founder with Darwin of the Theory of Evolution by Natural Selection requires that his awkward observations are integrated into the Theory and fully acknowledged.

PREFACE

ADDRESS DELIVERED TO THE THOREAU SOCIETY GATHERING AT CONCORD

Before I can address Thoreau's attitude to the state from the standpoint of today, we need to be quite clear as to his understanding of the condition of the human species as a whole. He determinedly avoided a merely abstract conception of the individual, the inexorable result of which would be an ideology. The nature of his fellow Concord citizens was not as straightforward or plain as may appear at first sight. In re-enacting the national "moment of origin" as Stanley Cavell puts it, Thoreau finds he has to re-define the nature of the species itself. What is the real state of the post-revolutionary American? How does he now differ by the mid-nineteenth century from the Old World denizen?

Early in *Walden* Thoreau remarks on the puzzling nature of the Concordians who "appeared to me to be doing penance in a thousand remarkable ways. What I have heard of Brahmins sitting exposed to four fires and looking into the face of the sun; or hanging suspended, with their heads downward, over flames; . . . or measuring with their bodies, like caterpillars, the breadth of vast empires; or standing on one leg on the tops of pillars—even these forms of conscious penance are hardly more incredible and astonishing than the scenes which I daily witness."

In twentieth century literature, the "hanging suspended" brings to mind the pupal stage of Kafka's Gregor Samsa in *The Metamorphosis* who enjoyed dangling from the ceiling. More directly, it recalls Thoreau's reflection in *A Week on the Concord & Merrimack Rivers* that "When I go into a museum, and see the mummies wrapped in their linen bandages, I see that the lives of men began to need reform as long ago as when they walked the earth." The act of mummification can be interpreted as a ritualistic imitation of the chrysalis stage of an insect's evolution, as Nabokov ironically projected in *Invitation to a Beheading.* There the criminal and writer Cincinnatus in that complained that his soul had "grown lazy and accustomed to its snug swaddling clothes," and while he is in that pupal state he feels he is surrounded by "wretched spectres, not people," by Apuleian larvae. Thoreau's Brahmins distinguish themselves from the house-dwellers of the Western world who are caught like "My gay butterfly entangled in a web," as *Walden* has it, or whose development has been arrested at some pupal stage like the British.

All this is not as remote from his *Reform Papers* as may appear at first sight, for at basis Thoreau is attacking the legal concept of 'person,' inextricably associated with the definition of 'property' in the judicial code, as well as with fixed, established 'perception.' Early on in *Walden*, he makes this quite evident: "I see young men, my townsmen, whose misfortune it is to have inherited farms, houses, barns, cattle, and farming tools; for these are more easily acquired than got rid of. Better if they had been born in the open pasture and suckled by a wolf, that they might have seen with clearer eyes what they were called to labor in."

Norman O. Brown, in an entirely different context, elucidates the process that unifies Thoreau's outlook in *Walden*. He sees a long extended structure of human maturation based on *insect* development determining underlying growth—

"*Larva* means mask; or ghost. *Larvatus*, masked, a personality—*larvatus prodeo* (Descartes); it also means mad, a case of demoniacal possession. *Larva* is also 'the immature form of animals characterized by metamorphosis'; in the grub state; before their transformation into a pupa or pupil; i.e., before their initiation."

There has already been a remarkable analysis some years ago of Thoreau's work viewing him through the prism of Brown in Joel Porte's *Emerson and Thoreau: transcendentalists in conflict.* But there Porte uses Brown's early book *Life Against Death.* If we draw as here on Brown's later *Love's Body*—which he once described to me as "indeed a strange episode, I have to continue burrowing underground, in strange directions"—then we draw very close to where Thoreau is trying to take us—and himself—in the parabola of his evolution. It is close to Charles Kraitsir's formula in his *Glossology* which Thoreau had read and where the "principle of motion, which preceded the birth of language" is embedded in early language roots following "the papillon of language, from the egg, through all metamorphoses." Accordingly to Kraitsir—and here I'm anticipating my later remarks on music—new words are added in line with the laws of the vernacular's "symphony."

But it is all too easy to be trapped in a larval state. At the end of *Walden*, Thoreau still has other lives to lead because he has only managed to shed so many skins, just as a caterpillar runs through at least 5 instars while in that preliminary condition. Thoreau's pupation has proved greatly difficult, and he drops back at the end to empirical perceptions, the gathering of natural facts in the hope once again that something will arise from these organically in terms of philosophy. Apuleius, master of metamorphoses, proposed a cluster of words in his *On the God of Socrates*:

"Now of these Lemures, the one who, undertaking the guardianship of his posterity, dwells in a house with propitious and tranquil influence, is called the 'familiar' Lar. But those who, having no fixed habitation of their own, are punished with vague wanderings, as with a kind of exile, on account of the evil deeds of their life, are usually called 'Larvae.'" So *larva* is not only 'person,' but one cast into the wilderness outside the bounds of established society, one of what Thoreau calls the "portionless, who struggle with no such inherited encumbrances." Someone in fact like the author himself, surveyor and saunterer, who emerges in time as a natural historian. And this relates to his search for a language sufficient for reality: "I fear chiefly lest my expression may not be *extra-vagrant* enough beyond the narrow limits of my daily experience, so as to be adequate to the truth of which I have been convinced . . ." For as the great theoretician of the science of language, Max Müller, wrote:

> "whatever view we take of the origin and dispersion of language,
> nothing new has ever been added to the substance of language,
> that all its changes have been changes of form, that no new root
> or radical has ever been invented by later generations . . ."

Philip Gura has observed in *The Wisdom of Words* that Thoreau demonstrates language is not a set of arbitrary signs but arises "organically from the very core of the empirical objects themselves, thus offering men profound clues to the organization of the universe." But how problematical Thoreau finds this is evident from the fact that the assertion early in his *Journal* that "Language is the most perfect work of art in the world" is transferred from humanity to the animals later in *A Week*. Maynard Keynes always held that animal high spirits motivated capitalist society. And Edgell Rickword saw that society as a merely animal society, not yet achieving human dimensions because thirled to the merely primal struggle for survival. An animal society is itself animated by what Thoreau calls "the most important part of animal [which] is its *anima*, its vital spirit." The human soul though is the ψμχ, psyche, imaged by the Greeks as a butterfly, so that insects take on the crucial role in defining the human. The author exclaims at one point in *Walden*—"sympathy with the fluttering alder and poplar leaves almost takes my breath away," and the poplar is etymologically related to papillon and to people, and was indeed during the French Revolution of 1789 when the poplar tree became a symbol of that revolution. Meanwhile Thoreau himself is trying to grow into a higher life: "in my solitude I have woven myself a silken web or *crysalis*, nymph-like, shall ere long burst forth a more perfect creature, fitted for a higher society . . ." The etymologies remote from the common coinage of speech take us back to an early flash of inspiration when in *The Natural History of Massachusetts*, Thoreau remarked "Entomology extends the limits of being in a new direction." This is a key also to Emerson's 1835 perception that language "thinks for us."

The great historian of the seventeenth century revolution in Britain was a close friend of Norman Brown—namely Christopher Hill—and the seventeenth century British revolution in a sense found its apotheosis and fulfilment in the American Revolution when the ancien regime was finally consigned to the proverbial dustbin of history. However as George Woodcock remarked: "the War of Independence had left Thoreau's fellow-countrymen both economically and morally enslaved." So although the externals of State organisation were altered, the psychological make-up was only partially modified, and it is this battle for a new and independent psychology that *Walden* enacts. The pioneering and revolutionary American character is at this stage, like Melville's Queequeg, "a creature in the transition state—neither caterpillar nor butterfly."

England meanwhile represents the form of domesticated atrophy par excellence, "an old gentleman who is travelling with a great deal of baggage, trumpery which has accumulated from long housekeeping, which he has not the courage to burn; great trunk, little trunk, bandbox and bundle." From this grows the moribund and oppressive nature of the British State which depends on its subjects' unwillingness to cast off the furniture of life as "*exuviae.*"

Interestingly Descartes, who is the source of Brown's original scheme of ontogenetic growth, had made explicit his own larval condition in his *Cogationes Privae* prior to his own emergence as a public philosopher:

"As an actor dons a mask, so that no one can see his own hue, so in the same way I who am going to take the stage before the world where I have been a spectator up to now, will be provisioned with a mask." Etymologically, a caterpillar or larva means a mask because its final form is concealed in its grub manifestation. The imago is a psyche which, as I said, the Greeks represented by lepidoptera. Brown goes on to remark in his *Love's Body* that "a person never owns [their] own person, but always represents another, by whom [they] are possessed. And the other that one is, is always one's ancestors; one's soul is not one's own, but daddy's." This is the meaning of the Oedipus Complex." So the foundation of the United States was both a flight from and confrontation with ancestors, and the seeking of an entirely new psychological as well as political make-up.

It is noticeable that it is the insects that draw Thoreau to Hindu thought: So in *A Week* he writes "the very locusts and crickets of a summers day are but later or earlier glosses on the *Dharma Sastra* of the Hindoos, and a continuation of the sacred code." Thoreau's identity matures within the systole and diastole of his natural surroundings. The fluidity of that identity allowed him, especially through his *Journal*, to transfer to the pulse of his writing the transitoriness of nature and its progressions. Like the bees he recorded which would arrive at the spring plants with unerring instinct prior to any possible observations of his own, he sought out nature in its formative metamorphoses, so that he could "make a chart of our life—know how its shores trend—that butterflies reappear and when—know why just this circle of creatures completes the world. Can I not by expectation affect the revolutions of nature—make a day to bring forth something new?" The rhetorical overstatement is part of Thoreau's strategy for diminishing and ironically deflating the human species. But it is also part of his strategy for defining how human life relates to nature, taking as his material his fellow Concord inhabitants and, above all of course, himself.

Some three years before *Walden* is published and before its final revisions, he provides a key to the 'plot,' so to speak, of his book:

"In the psychological world there are phenomena analogous to what zoologists call *alternate reproduction* in which it requires several generations unlike each other to produce the perfect animal—Some men's lives are but an aspiration—a yearning toward a higher state—and they are wholly misapprehended—until they are referred to or traced through all their metamorphoses. We cannot pronounce upon a man's intellectual & moral state until we foresee what metamorphosis it is preparing him for."

And extracts he copied into one of his Notebooks clarify his concept of self-development. He quotes James Elliot Cabot on "The Philosophy of the Ancient Hindoos"—"The essence of the Hindoo metaphysics, so far as they are of importance in the history of Philosophy, may be expressed in a few words: It is the reduction of all Reality to pure, abstract Thought."

In this connection, there is also a numerical, structural aspect: there are 4 Vedas, "the four works of supreme authority," 4 *asrámas* or stations of personal maturation, and there is "Brahma with four faces," while the four who keep house do so "according to four different modes." This is combined with various triplets of which the most significant is "the triple order of transmigration," where "'Souls endued with goodness, attain always the state of deities; those filled with ambitious passions, the condition of men; and those immersed in darkness, the nature of beasts: this is the triple order of migration.'" Although Thoreau's interest in Eastern thought faded after *Walden*, naturally in view of his emphatic and strengthening empiricism, his translation of "The Transmigration of the Seven Brahmans" demonstrates, as Arthur Verslius remarks, that he "had an abiding interest in making both Hindu and Buddhist works more available," and underlines "his interest in the doctrine of reincarnation and rebirth." In a sense, transmigration assumes an ecologically radical democracy, holding as it does the possibility that errant humans will have to pass through an animal or insect phase before salvation. Ultimately Thoreau's social equals are not the Concord citizens so much as its flora and fauna, for he is "no more lonely than a single mullein or dandelion in a pasture, or a bean leaf, or sorrel, or a horse-fly, or a humble-bee." This is Thoreau's natural democracy, what Lysander Spooner had earlier termed "the higher law than State law, natural justice."

Thoreau is not concerned with a simple born-again experience any more than with one-dimensional social adaptation, but with a multiple "moulting season," and leaves the woods ultimately because "I had several more lives to live, and could not spare any more time for that one.

" The interpretation of the allegory of human lives has its parallel in analysing and understanding literature and religion: "'They pretend,' as I hear, 'that the verses of Kabir have four different senses; illusion, spirit, intellect, and the exoteric doctrine of the Vedas;' but in this part of the world, it is considered a ground for complaint if a man's writings admit of more than one interpretation." Here Thoreau prepares us for the transformation of "a strong and beautiful bug" that will be resurrected at the end like some phoenix. This is the cornerstone of his strategy intended, as Robert Milder puts it, "to turn the unlikelihood of a human metamorphosis into a kind of fideistic certainty."

Throughout *Walden*, Thoreau is concerned with a humanity in transition, just as a climax in external nature is the onrush of spring, the sluicing out of winter. Indeed Emerson had theorized this as an aesthetic when he wrote that beauty lay in the "moment of transition, as if the form were ready to flow into other forms." Elsewhere Emerson raised the subtle complaint that machinery too often marginalizes intellect, and bids to reduce humans to "silkworms" and America to a "tent of caterpillars" (In "The American Scholar" he had contradictorily mused: "A strange process too, this, by which experience is converted into thought, as a mulberry leaf is converted into satin.") And this reflects ideas soon to be identified with Charles Darwin, but clearly already in the air, with regard to the inadequacy of rigidity in categories of defining species and genera. Later the American correspondent and advocate of Darwin, Asa Gray, will conclude: "This is a world of transition in more senses than is commonly thought . . . There are plants . . . which move spontaneously and freely around and among animals that are fixed and rooted." Thoreau is studying a humanity that, like the clay, takes on particular forms, characteristics, in response to the demands of social and economic changes, but is ever aspiring to transcend its condition. Emerson's over-soul becomes a type of under-soul, even undersoil.

In the seriousness of his quest, Thoreau often pauses to throw ironical light on his activities and although, or perhaps because, there is nothing like the book of Amadis of Gaul to guide him, there is every now and then a suspicion of Don Quixote in the extra-vagrant writer who declares "Be it life or death, we crave only reality." Stanley Cavell acutely observed in regard to Thoreau's remark "Our moulting season, like that of the fowls must be a crisis in our lives" that the use of "must be" suggests that "our moulting season, unlike that of the fowls, is not a *natural* crisis. Nature does not manage it for us. Our nature is to be overcome." From larva thru pupa to imago is a natural or given development; the person though has to maneuver him or herself through the stages by an act of will, faith or love. There is a suggestion, especially in the light of Thoreau's punning etymologies, that the "very ancient slough" (*W,* 6) of poverty early in the book has some link to this molting where "the snake casts its slough, and the caterpillar its wormy coat, by an internal industry and expansion." Comments Thoreau made to William Ellery Channing during their excursions around Concord are significant:

He spoke of the reserved meaning in the insect metamorphosis of the moth, painted like the summit sunrise, that makes its escape from a loathsome worm, and cheats the wintry shroud, its chrysalis. One sweet hour of spring, gazing into a grassy-bottomed pool, where the insect youth were disporting, the *gyrinæ* (boat flies) darting, and tadpoles beginning, like magazine writers, to drop their tails, he said: "Yes, I feel positive beyond a doubt, I *must* pass through *all* these conditions, one day and another; I must go the whole round of life, and come full circle."[28]

Indeed it is perhaps no coincidence that on the very day—April 28, 1856—that Thoreau "first definitely theorized the succession of forest trees," as Channing puts it, he was writing in his *Journal*: "As I was measuring, along the Marlboro' road, a fine blue-slate butterfly fluttered over the chain. Even its feeble strength was required to fetch the year about. How daring, even rash, Nature appears, who sends out butterflies early! Sardanapalus-like, she loves extremes and contrasts".[29] It is as if the developmental flashes of nature inspired Thoreau's inventiveness, measured his thought and flowed along the rivers of his consciousness.

To draw this now back to the natural sciences, Thoreau had read Coleridge's *Hints towards the Foundation of a More Comprehensive Theory of Life* when it first appeared in 1848, and one of the many passages he transcribed into his *Fact Book* held that "the insect world, taken at large, appears as an intenser life that has struggled itself loose and become emancipated from vegetation." Coleridge's *Hints* has been described as "a treatise on the use of natural history as means to the discovery of underlying laws of creation" and was an essential starting-point for Thoreau's fusion of imagination and ecological observation, although it has in recent years been down graded as an influence. What Coleridge crucially provided Thoreau with was the encouragement—even perhaps in terms of authorial authority, the *right*,—to merge a scientific approach to nature with his kaleidoscopic sweep over the living ecology of Walden. Let us recall that this is a time when the philosophy of Kant held sway, so that scepticism about the noumenon, the thing-in-itself beyond the phenomenon perceived by the observer, held sway. Emerson had of course coined the term Transcendentalism from Kant's categorization of intuitive thought as transcendental. Approaching matters through the prism of this New England intellectual milieu—even allowing for Emerson's awakening at the Jardin des Plantes—would in the first instance require some alternative philosophical engagement to satisfy both intuitive and practical perspectives. The rawness of the cry in *The Maine Woods,* which is contemporary with the writing of *Walden,* focusses his art on "the solid earth! the *actual world!*" down to the people themselves—"the *common sense! Contact! Contact!*"—in their demanding "*Who* are we? *where* are we?".

It is in this sense that Coleridge, speaking a philosophical language close to Thoreau's musings, would provide him with the impetus to deepen his investigations of the minutiae of nature. One of the key ideas of *Hints* is that if nature "had proceeded no further, yet the whole vegetable, together with the whole insect creation, would have formed within themselves an entire and independent system of life."[18] Thoreau's observations of insects developed soon after this, a slow gestation, a merging of his scrutiny of nature, and his reading. In the very year—1851—that his notes on flora and fauna became more specific and exacting, it is significant that entries in his *Journal* center on Ovid's *Metamorphoses.*

Zooming forward to a writer from Nabokov's youth, P.D. Uspensky, interestingly there are similar speculations in his *A New Model of the Universe.* Here Uspensky poses a genuine metaphysical-evolutionary riddle: "what place shall we give in this system to *insects,* which represent a world in themselves and a world not less complex than the world of vertebrates? May it not be supposed that insects represent another line in the work of Nature, and live not connected with the one which resulted in the creation of [humans], but perhaps preceding it? . . . Insects reveal, in their structure and in the structure of their separate parts and organs, forms which are often more perfect than those of [humans] or animals . . . At some point, ants and bees . . . lost their ability to evolve and after this Nature had to take her own measures and, after isolating them in a certain way, to begin a new experiment."

Thoreau's comments on music are often overlooked, but although they are early comments they dovetail perfectly with his lifelong projects. In "Slavery in Massachusetts," he connects discordancy with injustice and complains in "The Service" that our lives are "full of abruptness and angulosity" lacking in majesty, and he surmises "a world of peace and love" where "music would be the universal language, and men greet each other in the fields in such accents, as a Beethoven now utters at rare intervals from a distance. All things obey music as they obey virtue . . . When we listen to it we are so wise that we need not to know."

Now, the language elements we've been considering constitute among the insects a quaternal structure from the ovum through the caterpillar to the pupa and finally the imago. Beethoven's symphonies naturally follow this shape: In the 1st movement main themes are stated; the 2nd or slow movement proceeds like a caterpillar, it crawls; the third is febrile and anticipatory like a pupa twitching for realization, or shimmering chrysalis (think the scherzo of the Eroica) while the 4th is the climax, as Berlioz analysed it—"from tension to release, from compulsion to liberation, from the tragic to the joyous." (I give a more extended analysis of the music into insect, insect into music in my chapter on Schopenhauer in the book *The Insect-Populated Mind*). Music Thoreau's "science of melody and harmony" built on mathematical structures as he describes it in *A Week on the Concord & Merrimack*, "is the sound of the universal laws promulgated. It is the only assured tone." And again in *The Week*, "the most distinct and beautiful statement of any truth must take at the last the mathematical form. We might so simplify the rules of moral philosophy, as well as of arithmetic, that one formula would express them both."

And if we follow the Hindu Sanyasas which he mentions, again there is a fourfold cone that imitates insect growth from the brahmacarya of education thru the garhasthya stage of the active citizen to the retreat for the loosening of bonds to the sannyasa or life of the hermit. Edward Dahlberg defined *Walden* as taking "its inspiration from the *Vedas*" to be "the secular bible of our ethics." For Hinduism is not the acceptance of academic abstractions or celebration of ceremonies, but a mode of life and experience. It is insight into the nature of reality (darsana) or experience of reality (anubhava). This experience is not an emotional matter but a response of the whole personality, the integrated self to the central reality. Contemplation rather than other forms of thinking is the essence. And we might also remember what Heidegger said—dubious witness though he proved to be in his life-decisions—"we haven't got a world, any more than a stone has, unless it is a fourfold one and it says something about us that we are bound to find this a preposterous statement. But it is only with respect to the preposterous fourfold that we can arrive at any sense of what it is for language to speak."

So how does all this relate to the transformation of that leviathan, the State? Melville of course created the epic of the unforgiving and unremitting nature of the state in *Moby-Dick*; governing States are merely force concentrated in the hands of a few. Gandhi, inspired as he was by Thoreau believed "we must become the change we want to see." And this has been the philosophy of the movement for Scottish independence; as the writer and painter Alasdair Gray expressed it, one must go on each day as if the nation were already independent, the idea being that the constitutional situation would be forced to fall into line with the behaviour of the living members of the society in the course of time. And that is exactly what is happening, with the Westminster government of Brown having to reluctantly follow the localized advances made.

On the international scale, there is a direct relationship between Thoreau's opposition to the war with Mexico and the invasion of Iraq. For the ideology of 'Manifest Destiny' was coined by a journalist in 1845 immediately prior to the Mexican war, and it is this disastrous concept that lies behind the multiple errors of the Iraq venture, the utopian disbandment of the apparatus of the army and of the localized Ba-athist official structure. As I wrote in a British newspaper recently: "the cultural devastation of Iraq called for by the prophet of the Babylonian exile, Ezekiel—"The end is come upon the four corners of the land."—may have taken two and half millennia to come to pass, but the White House with its annex at Number 10 Downing Street have accomplished the mission. Thoreau wrote in *Resistance to Civil Government*: "How does it become a man to behave toward the American government today? I answer, that he cannot without disgust be associated with it." So the tax-resisters in the States (there are none to my knowledge in Britain) follow in his footsteps in assuming: "There is a higher law than civil law—the law of conscience," which echoes Lysander Spooner's insistence that natural justice supercedes State legislation and the impositions of State jurisprudence. *Walden* unfurls a unique plan of regeneration for Western culture, and it is one that cuts athwart the assumptions of State control and manipulation.

The merging of inner and outer worlds, psychological and political transformations, prepares the way for what my old sparring partner Norman Brown insisted was the necessary apocalypse. The final book of his great trilogy on civilization and its aspirations is, after all, entitled *Apocalypse and/or Metamorphosis*. And I should like to end with some wise words from William Zinsser addressed to members of the Authors Guild:

"the truth no longer matters to the people in charge of our lives. The White House has repeatedly lied to us; corporations repeatedly lie to their stockholders by manipulating their figures, which enables them to ransack the companies and their pension funds . . . .[An] oddity is that not many people seem to mind. Where are the cries of outrage from the traditional guardians of moral authority? Where are the ministers, where are the bishops, where are the rabbis, where are the college presidents [to express] not only anger but sorrow that the fundamental decencies of this country are being eroded?"

And I might add that things have just got worse. Truth itself has been abolished and been replaced by Hillary Rodham Clinton's term—"mis-speaking."

CHAPTER I

THE INVENTION OF LANGUAGE EMBODYING ITS DEFINITION OF HUMAN NATURE

The greatest twentieth century intellectual practicing politician—André Malraux—wondered whether a new attempt to understand humanity would be undertaken "when the awareness of metamorphosis comes to dominate evolution and to create its own history."[1] Since then Norman O. Brown has lit upon a key set of language elements via the insights of Descartes:

Larva means mask; or ghost. Larvatus, masked, a personality—larvatus prodeo it also means mad, a case of demoniacal possession. Larva is also 'the immature form of animals characterized by metamorphosis'; in the grub state; before their transformation into a pupa, or pupil; i.e. before their initiation.[2]

So the butterfly conceals its final form at all stages;

ovum (egg) > larva (caterpillar)> pupa (in its chrysalis: from crusos {gold}) >

imago (perfect final creature)

This is a fourfold process and why this is so important will become clear as we progress. In the course of the book, we shall be seeing how this structure in nature has been appropriated by some of the greatest poets from Blake to Coleridge and T.S. Eliot. Quite beyond merely descriptive or passionate expression, the writer threads into the natural world at his and her heart. The integration between the writers and that structure meets its point of fusion in their representation of insects. The dialectic of microcosm (the poems) and macrocosm (the dimensions of the universe) is in play throughout.

In the wake of the French Revolution, the German writer Novalis posed a number of issues about language and consciousness that still await answers. He was doubtful as to the virtue of the academic mind. Such an orderly trained intellect "goes in swiftly—but also comes out swiftly—He soon reaches the second stage [i.e., in our interpretation, the larva]—but there he usually remains. He finds the final steps difficult and, once he has attained a certain level of mastery, can seldom bring himself to revert to the condition of a beginner."[3] Now what are these stages of intellectual progress Novalis is referring to, and why should this academic mentality need to return to the Yeatsian "foul rag and bone shop of the heart" ('Circus Animals' Desertion') before proceeding to the third stage? Because there has to be a complete re-learning process undergone, just as in insects the cells that defined the caterpillar are destroyed in order to release the cellular pulp that is refashioned and hardens into the pupa. And Cassirer noted the tradition shared by Milton and Coleridge of "Platonic thought that describes consciousness as an evolving, multi-stage process that is akin to conception, gestation, delivery, and birth." For Novalis, it is ordinary people who are best equipped to tackle the higher echelons of development. The "confused," as he calls them

penetrate slowly, they learn to work with difficulty, but then they become masters and teachers for ever . . . Confusedness indicates superfluity of strength and powers, but lack the sense of proportion. Precision—and true sense of proportion, but scanty strength and power. That is why the confused man is so perfectable (sic) compared with the trained man who so soon finishes as a Pedagogue.[4]

So the 4 language elements reveal a tale of insect evolution, running in allegorical parallel to human intellectual and spiritual maturation. This is the metaphor that is humanity, and it coincides with the "peculiar property of language" which as Novalis says in his Monologue, "is concerned only with itself."[5] But neither he nor his predecessor Fichte can reveal it in detail as the "truly scientific" original language. Not only was language when fully revealed Delphi itself, the route to self-knowledge, but also "the dynamic element in the physical realm."[6] Language appears to be a mere utility. We use it as common coinage for every conversation. And in the world of politics and economics as Novalis shrewdly notes, "where there are many words, there must also be much activity—as with the flow of money."[7] This is on the same cultural lines as Keynes' percept that capital is generated by animal spirits in human action.

Max Müller has written:

Roots may seem dry things as compared with the poetry of Goethe. Yet there is something more truly wonderful in a root than in all the lyrics of the world.[8]

It is a truth whose significance has been too rarely understood that no new root has been invented since the Ancients. Certain words and roots are every bit as central to our understanding of ourselves and our species as DNA and Darwinian evolution. Indeed they are simply a different branch of our evolution, locked into our cultural and spiritual history. There is a defining message in language, with natural selection at work linking crux language elements. Or as Müller put it—"there is a petrified philosophy in language."[9]

In the early days of humanity, sounds were instinctively articulated to express the core of objects confronting human experience. The objects spoke through the individual response revealing their essence. This process is in accord with the prevailing mode of thought in Homer. There, the psyche does not belong to the person. When Achilles reacts, it is not as an individual character; he is simply penetrated by external forces to the core of his being. The soul is purely an echo of the objective world. The Greeks imagined it as a butterfly (ψυχή). However "the creative faculty which gave to each [thing] a general conception, as it thrilled for the first time through the brain, a phonetic expression, became extinct when its object was fulfilled."[10] After the childhood of humanity, immediately natural things were put aside and the faculties associated with them attenuated and withered.

In time these sounds were enshrined in the Indo-European roots that determine all major languages. As the child learns language almost instinctively via parents and teachers, so the four or five hundred roots common to all widely used languages came into being through a phonetic power. Noam Chomsky has argued we are hard-wired for grammar; but by the same token there is a creative faculty for articulating roots that I will show enables a deeper understanding of human nature. As Müller declared:

Things awakened within the individual sounds that materialized as roots and engendered phonetic types on the basis of which a body of language was progressively formed. However once human beings ceased to 'resonate' before the world, a sickness invaded language. Humans became the victims of illusions produced by words.[11]

That is to say once the youth of humanity passed away long before the arrival of the Buddha, Christ or Muhammad, language had become common coinage in which the root meanings of words were obscured, except in the work of the poets and philologists.

And so language became the common currency of communication rather than "a petrified philosophy" to be animated and its message interpreted.

To bring this elemental philosophy to life is the aim of this book. And in its course, I hope to show that the obsession with the evolution of the human species from the great apes by way of natural selection overlooks evidence far more intimate to our development. Language tells us we are equally related to insects, unpalatable as this may seem. The physiological and biological evidence is on one level obvious. Our coccyx reminds us of our simian origins, and the dip in our body temperature in the middle of the night recalls the tree shrews we inherited this feature from. Our relation to the insects has to be dug out from the whole fund of our scientific and cultural knowledge.

Here though we are following a trail that leads away from contemporary language to Ancient Greek and Sanskrit. Here we find the roots of present-day words, and it is a linguistic truth that no new root has been invented since the ancients. Certain words and roots are every bit as central to our understanding of ourselves as DNA and Darwinian evolution. There is a defining message in language, and the natural selection here is the interlinking and survival of the necessary language elements. In early days of humanity, sounds were instinctively articulated to express the core of the objects being experienced. The objects spoke through the individual response. These sounds were then enshrined in the Indo-European roots that determine all major languages. As the child learns language almost instinctively via parents and teachers, so the four or five hundred roots common to all languages came into being through a phonetic power. Certain 'animal' structures inhere in the embryonic mind and are there in the ovum. The clues lie in strings of words conceived during the very birth of language, and apprehended before cultures had ossified into the Greek and Sanskrit mainstream, a type of molten big bang in the language world. It is not, pace Chomsky, that the embryos are forming for sentence structures. It is that as the Scottish poet Hugh MacDiarmid put it: "There lie hidden in language elements that effectively combined/ Can utterly change the nature of man."[12]

Charles Darwin recognized in The Descent of Man that:

> the mental powers in some early progenitor of man must have been more highly developed than in any existing ape, before even the most imperfect form of speech could have come into use; but we may confidently believe that the continued use and advancement of this power would have reacted on the mind itself, by enabling and encouraging it to carry on long trains of thought. A complex train of thought can no more be carried on without the aid of words, whether spoken or silent, than a long calculation without the use of figures or algebra.[13]

It was work in all probability that gave rise to the sounds arising from differences in breathing lengths associated with specific labours. So dar- is a root meaning 'tearing,' and by modification in the course of time enables the word 'tree,' itself still close to the word 'tear,' which itself perhaps has its own lachrymose implication. This came via the Greek root δείριο and Sanskrit dru, thence to the Greek δρῦς (dryad). The idea behind this association was that tearing the bark off trees and hollowing out the trunk, kayaks, tables and other things could be created. Hence at the same time a concept was born. Roots are like the fonts in the traditional printer's tray. They are the shapes in which all words are cast. A primary root leading to the word 'earth' is ar- to plough. This appears in Shakespeare's sentence "to ear the land that has some hope to grow." Latin then produces the name of a plough from it, ara-trum, while it will also become ars, art. 'Article' also emerges from ar-, as a literal translation of the Greek αρθρώ, meaning the socket of a joint. So for Aristotle who first used the word 'article,' it meant the common words which formed the sockets in which the members of a sentence were contained. As Franz Bopp first remarked, the broad outlines of grammar in Sanskrit, Greek, Latin and Gothic are almost identical without any mutual influence. Apparent differences are explicable though phonetic corruption and the speech idiosyncrasies of each nation. Greek grammar owed its origin like the Greek language itself to the critical study of Homer, while Sanskrit arose from the study of the ancient Vedas. The similarities of words in the two languages of Sanskrit and Latin suggest they are what Müller calls "relics of the primitive language of mankind."[14]

So we have had to proceed to the roots of key words and to a helical series of words—to paroles rather than langue in Saussure's and Barthes' terms—to cast further light on the human-insect connection. My theory takes an earlier starting point than Chomsky's Universal Grammar and Terrence Deacon's grammatical intent. It is directed at language elements, even the root elements of language itself. It takes words that clearly arose when humans were still extremely close to organic nature and its processes. In that sense, it depends on a relatively early and simple function of the brain, little affected by the mediating prefrontal cortex. As William H. Calvin puts it: "rather than a gene for a language machine, you might have an epigenetic [cellular pathways] tendency to seek hidden patterns in your sensory environment." Chomsky, like William Calvin, is focused always on syntax of which his deep structures are the manifestation. In his *Extraterritorial*, George Steiner though expressed his reservations about Chomsky's perspectives: "I disagree with Chomsky's dismissive ruling on the relations between linguistics and certain aspects of biological and evolutionary theory . . . A theory of the innateness and generation of language in man which has no substantive regard to the biological, evolutionary, social aspects of the phenomenon will remain necessarily arbitrary and incomplete."[15] And elsewhere Steiner argues that in language "we are unquestionably aware of a constant movement towards immateriality, a process of metamorphosis from the phonetic to the spiritual." This is exactly my point, except that having moved from the phonetic to the spiritual, one then has to re-interpret the spiritual in entomological and evolutionary terms. The paradoxical fusion of ape and insect in the human condition is the pivot.

Writes Terrence Deacon in his splendid book, The *Symbolic Species*: "In some ways it is helpful to imagine language as an independent life form that colonizes and parasitizes human brains, using them to reproduce."[16] While he goes on to modify his definition of language as a virus, I believe it perfectly describes the fragmentary evolution of key words. Deacon prefers to define language as "a social phenomenon,"[17] and while there can be little doubt that communication at the time of hunting played its part, this is not the primary nature of language. Though the prefrontal cortex is the pivot for more sophisticated mental activities, it is not the key to the ability to understand words and even sentences, as those whose cortex is damaged show. Also in word association, so crucial for poetry, other and more primitive areas of the brain are involved as well as the ventral prefrontal cortex. Here the contralateral cerebellum comes into play. There is a wide range of language processes at work throughout the brain, and no evidence of a language module. To a degree, that is reflected in the fact that, as Schopenhauer argues, we don't think in concepts but in their representatives—words. Or as Plato wrote—in logos, in images of thought.

Moving on to contemporary writing, Julian Ríos in his Joycean epic of etymologised experience, *Larva*: Babel de una Noche de San Juan (Larva: Midsummer Night's Babel), surmises with Cabrera Infante, that the Spanish language is "a larval mask beneath which other languages lurk."[18] The whole enterprise of literature is projected as a vast larval undertaking in which these "other languages," or simply another perhaps trans-human, language is striving to emerge. At the plainest level, it is interesting that Spanish has a number of words of entomological exactitude quite lacking in English such as 'el horadado,' which is a cocoon bored through—and specifically a silk-worm's cocoon. But Ríos is clearly not thinking of this line of consideration. He is rather hinting at a revolution involving the issues formulated by Octavio Paz, now being raised by a novelist as, with the exception of Valente, the poets have largely abandoned this concern: "touched by poetry, language is more completely language and, simultaneously, it ceases to be language; it is a poem. An object made up of words, the poem opens into a region that is accessible to words." The "other languages"—not only root languages per se—are of a different order from the word. Again Paz makes wonderfully exact such possibilities: "In order to transcend something it is necessary to pass through that something and go beyond it: music does not transcend the articulated language because it doesn't pass through it."[19]

If "all art aspires to the condition of music," it can never reach it by merely imitating it, as writers like Robert Bridges prove in their failures. Any identity between the arts is embodied in their structures, and as opposed to Kantian a priori forms imposed on, or given in the poet's mind, the forms arise from the natural world and take shape in language. But these processes are not directly related to our animal ancestry. They are obliquely present in our relationship to the insect world and its modes of growth.

Modern Spanish and Spanish American writers have generally rejected the legacy of logical positivism as delineating over-narrowly and one-sidedly what is objective. Essentially a new philosophy is implicit in their work, even if no-one except Paz and, to a lesser degree, Aleixandre and Carrera Andrade, close in on the conclusions. (see my book The Poem and the Insect: aspects of twentieth century Hispanic culture). The crux lies in that area of the part played by language in human thought and experience. It was Wittgenstein, of course, who put language at the core of twentieth century philosophical concerns, and there is no indication that the twenty first century will dawn with more than a few logic and methodological points decided. If Wittgenstein asked all the key questions then, as is well known, he failed to elaborate a general theory of language in the *Tractatus*, and ultimately abandoned the idea of such a theory. He starts off with an outlook towards what Chomsky will call "the language faculty" similar to that of the American thinker, but he never goes so far as hypothesizing a Universal Grammar as a "theory of our biological endowment."[20] However in his remarks on the limitations of a private language, Wittgenstein argued that "an 'inner process' stands in need of outward criteria."[21] In many ways, this is what poets as disparate as Vallejo and Aleixandre always sought. The commitment to this search of the poets up to the Civil War was, in its own way, heroic and certainly single-minded. Later Wittgenstein abandoned the quest, arguing for a purely social and utilitarian value to language, the meaning of a word—one should rather say the significance of a word—lying in the aggregation of its uses within specific statements. He concludes that language is a game.

A return to the roots of language, though, brings into play both necessity and spontaneity, so that many clusters of Greek, Latin and Sanskrit words were, in their evolution, responses to the actual world in the light of relatively fresh experience. In that sense, Hugh MacDiarmid's visionary insight quoted earlier in this chapter can most realistically be further illuminated by reference to an early stage of the development of language, and centres on discrete words ("elements") as opposed to sentences. To get at these elements the poet has to turn him or herself inside out, as it were, and try for the improbable by getting to the other side of language in order to look anew at the interface of the verbal in relation to reality. Wittgenstein pronounced this impossible, but it is this that so many Hispanic poets, especially the Spanish-American with their more marginal geographical position and outlook towards the languages and powerbase of Europe, essayed. However they grasped that the image short circuits the idea that all experience and thought has to be linguistic, has to take the form of linguistic articulation. It is true that Wittgenstein maintained that sentences picture and are instruments, while words denote and are tools. But the structure of sentences is a consequence of the nature of human activity, whereas the assemblage of interlinked groups of words, especially in their epistemological origins, goes beyond the functional. Frege's insistence that the sentence is the monad, rather than the word, is likewise flawed. Undoubtedly the (often residual) continuing power of the

Bible in Hispanic society has much to do with the potency, the positive energy and even the magic of the Word. The quest for significance, for a profounder balance between subject and object, takes place within the theoretically problematical arenas of literature, music and religion instead of philosophy as such, because the answer is sensed to be in those spheres. Of all the great twentieth century literatures, the Hispanic has laid bare, dramatized, and even enacted in the lives of its poets this struggle and this search.

Insects can metamorphose, destroying their former selves and building a new stage of evolution. It is a sleight of nature, an unusual evolutionary mechanism. And it is built into the structure of our languages. When Descartes came out of the shadows where he had been working on his theories, he sensed a parallel with his own development. Larvatus prodeo, he announced—"I advance masked:" And Novalis, whose ideas fed into Emerson, Thoreau and the Transcendentalists, sees nature as a series of ciphers to be read and interpreted, correspondences between the world of nature and the human mind that can activate true knowledge. Metamorphosis itself, of course is far older than its discovery. It was in the Devonian period, 350 million years ago, that it developed in the amphibians. Still today, tailed amphibians and anurans (toads, frogs) pass through a larval stage before they reach maturity. It was also in the epoch 300mya that the FOXP2 gene formed, a determinant of language ability only recently discovered. It is the insects, diverging from a common arthropod ancestor 300-400 million years ago, that are the principal species growing to adulthood through a series of transformations. In evolution, the roots of this remarkable process are relatively straightforward. Metamorphosis occurs because the larval form, the pupal form and the adult form have evolved independently of each other to fit different environments.

One of the confusions in thinking about language, as opposed to instinctively using it in everyday conversation, is that in our time there are two separate words for language and for reason. The ancients had one— logos. The significance of life was believed to be sealed into language. In the Bible this logos is identified with God in the St. John Gospel: "In the beginning was the Word, and the Word was with God, and the Word was God." This Word though is no ordinary word. And according to one of the main Christian theoreticians it turns out to be no actual word at all, so urgent is the need to mystify. Augustine of Hippo rather fancifully defined it as "anterior to every sound and to every thought of sound . . . This is the word that belongs to no language, the true word about a true thing, having nothing from itself, but everything from that knowledge from which it is born." So paradoxically this Word is pre-linguistic. Theologically, the Son is the Holy Trinity, is the Word, a person then who is no person but a concept, offspring of God's intellect. Which approximates to Foucault's concept of language where, following Blanchot, it is "only a formless rumbling, a streaming; its power resides in its dissimulation. That is why it is one with the erosion of time; it is depthless forgetting and the transparent emptiness of waiting." And likewise in much modernist thought, including Beckett, language is an aspiration to silence, "a progression towards silence, the absolute speech." However for the Holy Family, the Son is the Word, a person who is not person but a concept, offspring of God's intellect.

In an attempt to move beyond this sophistry, let us begin with the most ancient word for 'butterfly,' aka 'soul'. This word is φαλλαινα. This appears in Nicander, and probably refers in the first instance to a moth, one of the Noctuids, Agrotis pronuba—and to judge from the pronuba element in its definition, an Orange Underwing originally. The Greek word itself is related to φαλλός, phallus, which is also the word for 'whale.' Both derive from the root *bhel, 'to swell up.' The earliest depiction of the butterfly as representative of the soul is on a black-figured amphora of the sixth century BC. Here the insect is receiving drops of semen emanating from a satyr, while a naked female figure is dancing in front of him. As Carl Schlam puts it: "This juxtaposition can be read as an expression of the living force about to join an embryonic body."[22] Significantly the potential body is in fact a lepidopteron. The connection with the whale is probably via the sperm-whale, and brings to mind Herman Melville's prescient Moby-Dick, itself of course a sperm whale giving Captain Ahab a hard time. Perhaps there is even a sly pun on Melville's part in the 'Dick' component of the name!

According to Otto Immisch—a true master at the turn of the twentieth century at winkling out the significances of Greek scripts—Nicander saw elements of a rudimentary symbolism in φαλλαινα that failed to reach the level of created conception embodied in ψυχή.[23] This later word, with its breathing sounds, comes to express in the course of time, breath itself. Tecumseh Fitch has speculated that "early proto-language may have contained just a few isolated words and very little grammar," and "protolanguage was made up of individual words unconnected by complex syntax."[24] As Eric Havelock wrote: "The Greeks dissolved the syllable into its acoustic components—we might almost say its biological component in as far as these are actually effects produced by the movements of different parts of the human body."[25] The psyche in Homer leaves through the mouth; it is breathed forth and flies from the body at death for its journey to Hades. There, as Bruno Snell relates, "it leads a ghostlike existence as the specter (ειδολον) of the deceased."[26]

CHAPTER 2

TRANSFORMATIONS

The Scottish visionary Patrick Geddes lived for a time in the house once occupied by the prototype for Stevenson's Dr. Jekyll and Mr. Hyde. He looked on it, so he told his biographer, as an imaginative if ambiguous commitment, positioning him strategically as it did for his raids on the University:

Long ago, I bought this fine old house in High Street and carry on the business. I am a burglar by profession too . . . That's my secret! My diagrams are really skeleton keys, and to ever so many of my colleagues' departments of sciences, philosophies and what not, so I go round by day and burgle more universities than this one.[1]

His method is not dissimilar from that of any thinker today ransacking all intellectual fields to gather the theoretical coordinates able to counteract that fearful paralysis diagnosed as mummification by Nabokov's Cincinnatus in Invitation to a Beheading. This criminal and writer complained that his soul had "grown lazy and accustomed to its snug swaddling clothes," and while he is in this pupal state he feels he is surrounded by "wretched spectres, not people," semi-created outcasts on the lines of Apuleius' larvae or the unmade of William Blake.[2]

So the spine of my book is fourfold metamorphic, and constitutes a journey inside an almost genetic helix. I shall be arguing that natural processes of transformation have their reflection in the greatest cultural achievements of the human species. And to follow the arguments of this book, there are a number of paradigms already sketched in Chapter 1 that the reader needs to keep in mind:

❖ the processes of insect development: ovum, egg → larva, caterpillar → pupa, chrysalis → imago, fully grown insect.

❖ the structure of sonatas and symphonies: 1st movement, statement of themes (ovum) → 2nd, typically a slow movement (larva) → 3rd, scherzo, febrile pace (chrysalis) → 4th, finale, culmination (imago).

❖ the organic growth of the personality: infant (ovum) → child (caterpillar) → early adult (pupa) → mature adult (imago).

This is the story of how the world of insects has penetrated to the innermost reaches of human experience. It involves music, language and genetics. As in the world of physics, what we observe as the commonsensical world is only a fraction of reality. Quantum data is an even larger part of what is going on out there. The formulation of a theory that unifies the Einstein model and the small and often infinitesimal quantum atomic activity remains to be discovered for a fuller version of the real world. Such a theory will undoubtedly enhance our appreciation and understanding of life. This book aims to show how there is also a parallel quantum world of natural organic history, that of the insects, that also needs to be unified into a fuller evolutionary apprehension of their links with the great apes and thus with humans. By now a century and a half after Darwin's publication of The origin of species, every schoolchild knows we are directly evolved from apes. But because of the limited state of genetics, embryology and morphology, Darwin could not be expected to distinguish the equal and, perhaps even more significant, relation with the insects. However when evolutionary biology met embryology and genetics, it opened up an entirely new perspective on the profiles of species.

Richard Dawkins has splendidly spelt out the remarkable change over the past 25 years of research:

It was a similar triumph to show that the insect head contains—again all jumbled up—the first six segments of what, in their remote ancestors, would have been a train of modules just like the rest of the body. It was a triumph of late twentieth century embryology and genetics to show that insect segmentation, far from being independent of each other as I was taught, are actually mediated by parallel sets of genes, the so-called hox genes, which are recognizably similar in insects and vertebrates and many other animals, and that genes are even laid out in the correct serial order in the chromosomes! That is something none of my teachers would have dreamed of when I was an undergraduate learning, entirely separately, about insect and vertebrate segmentation. Animals of different phyla (for example, insects and vertebrates) are much more united than we ever used to think. And that, too, is because of shared ancestry. The hox plan was already sketched out in the grand ancestor of all bilaterally symmetrical animals. All animals are much closer cousins to each other than we used to think.[3]

The situation began to change some quarter century ago with the discovery of the homeobox set of genes, and after this, the main features of animal development were found through Drosophila merganogaster. Further research has shown how between 120,000 and 30,000 year ago, "the effusive mass of tangled neurons took a significant new step that formed the essential patterns of thought and greatly increased human consciousness . . ."[4] And the recent discovery of the spindle cells has thrown more light on the evolution of brain and consciousness. These cells appeared some 10-15 million years ago in the common ancestor of humans yet to be discovered, and rapidly increased 100,000 years ago. The spindle cells deal with high-level emotions and exist in the fronto-insular cortex.[5]

At the same time remarkable advances have been made in understanding the nature of the brain stem, which has been inherited from amphibians and reptiles. The stem lying at the boundary of the brain and spinal cord largely controls our breathing, heart-beat, digestion and sexual drive, everything involuntary in our lives. As Neil Shubin writes: "the brain stem originally controlled breathing in fish; it has been jerry-rigged to work in mammals." It is Shubin who has identified hiccups as the clue to our debt not only to amphibians in general, but specifically to tadpoles. The sudden closure of the tadpole's glottis to prevent water entering the lungs is the ancestral experience that explains the advent of hiccups. Shubin's Arctic researches into fossils produced the clues for our relationship to aspects of amphibian physiology. He discovered the transitional water-to-land creature, the Tiktaalik, "which is just as much a part of our history as the African hominids, such as Australopithecus afarensis, the famous 'Lucy.' Seeing Lucy we can understand our history as highly advanced primates. Seeing Tiktaalik is seeing our history as fish."[6] Indeed recently in 2012 (reported in Biological Reviews) a prehistoric eel-like creature has been discovered in a Canadian shale bed and been identified as the earliest ancestor of humanity

So our heritage is very mixed. I hope to show by the end of this book how it is even more mixed when you take into account the natural reflexes of the working of the human brain, and its evolution through cultural achievements into the human mind. Yet the overall biological span travels from tadpole or eel to butterfly.

CHAPTER 3

THE CLASSICAL WORLD BECOMES QUANTUM

Modern philosophy kicks off with Descartes. As he emerged from obscurity, he famously declared in 1619 "larvatus prodeo," "I proceed masked." He goes on to say in this early manuscript: "The sciences now have masks upon them: if the masks were removed they would show up as extremely beautiful."[1] He has a "contempt for logic," and from the first his thinking takes on a double–helical character. The modern philosopher was to be a new kind of human. Hobbes had already picked out the linguistic associations involved in early modern self-definition in his all-encompassing Leviathan:

The word Person is latine: insteed whereof the Greeks have πρόσωπον [prosopon], which signifies the Face, as Persona in latine signifies the disguise, or outward appearance of a man, counterfeited on the Stage; and sometimes more particularly that part of it, which disguiseth the face, as a Mask or Visard: And from the Stage, hath been translated to any Representer of speech and action, as well in Tribunalls, as Theaters. So that a Person, is the same that an Actor is, both on the Stage and in common Conversation; and to Personate, is to Act, or Represent himself, or an other.[2] Norman O. Brown picked up the masking phrase from Descartes' Cogitationes Privatæ (1619) and from Hobbes, transforming it into his stunning sequence earlier quoted.

Brown himself elaborates on his theory of mask, via Erving Goffman.

Personality is persona, a mask. The world is a stage, the self a theatrical creation:

The self, then, as a performed character is not an organic thing that has a specific location, whose fundamental fate is to be born, to mature, to die: it is a dramatic effect arising from a scene that is presented." The self does not belong to its possessor. "He and his body merely provide the peg on which something of a collaborative manufacture will be hung for a time[3]

Imago, of course, is the perfected insect, the butterfly itself or moth, but also the image, what the Greeks interpreted as a wraith of the dead person's unquiet soul, or eidolon. The magisterial philosopher of the Psyche, Erwin Rohde, drawing on the evidence from Homer put the matter like this:

According to the Homeric view, human beings exist twice over: once as an outward and visible shape, and again as an invisible "image" which only gains its freedom in death. This, and nothing else, is the Psyche.

Such an idea—that the psyche should dwell within the living and fully conscious personality, like an alien and a stranger, a feebler double of the man, as his "other self"—this may well seem very strange to us . . . It was the experience of an apparent double of the self in dreaming, swoons, and ecstasy, that gave rise to the inference of a two-fold principle of life in man, and of the existence of an independent, separable "second self" dwelling within the visible self of daily life.[4]

These parallel existences within a single individual were abstracted by Descartes, and molded into a whole philosophy. Metaphysics is the root, while the trunk is physics and then there are the branches of knowledge with the fruit at the end of a branch. Metaphysics and physics are equal aspects of philosophy and both are sciences, or scientia, knowledge in the strictest sense. For Descartes, the soul is a substance distinct from the body, and it is a rational entity as opposed to the Aristotelian "vegetative and motive force of the body" which is shared with plants and animals. Animals, according to Descartes, have no anima, which applied only to man. By 1636 he had already outlined a project for a universal science to raise our nature to the highest degree of perfection, demonstrating the separation of the soul from the body.

Dieter Heinrich has advocated the bringing to completion of philosophical ideas where the author has not been fully aware of the potential advances possible by way of his methods. I believe this to be the case with Descartes.[5] The finest writer on Descartes, Stephen Gaukroger, argues that "any attempt to show that the soul must both have an existence independent of the body and an identity independent of the body can only be dualist." But if the psyche is both a reflection of the natural world (as in the entomological progression to imago), while presenting itself to the unknowing human as a spiritual or abstract entity, then mind and body become a single object. In the light of advances in knowledge over the past 40 years, we can reconsider the volleys ranged against Descartes' dualism.

In his Preface to Principles of Philosophy, he argues that the human soul is immortal and separable from the body. And he believes that true—which is to say metaphysical—knowledge is free of the evidence of the senses, "even somehow opposed to what the senses habitually conceive. This is entirely the situation with the quantum world, a sub-atomic region which is not susceptible to common sense(s). So Descartes is most succinctly characterized as saying 'I doubt and know I doubt. Therefore I am.' The principles of metaphysics, as Descartes states them in the preface to the Principles of Philosophy, are that God is the creator of all beings and the source of all truth, and that the human soul is immortal and separable from the body.

His principles do not have their origin in the scholastic metaphysics of Aristotle but in the Augustine doctrines of God and soul. Descartes presented Augustinian thought as suitable first principles for philosophy. He puts himself forward as the philosophical proponent of the Counter-Reformation against Aristotle:

Christian philosophers had felt bound to maintain the doctrine of the separability of the human soul, which Augustine (following the Platonists and the Greek Fathers) had made normative for Latin Christianity; and Thomas Aquinas had managed to interpret Aristotelian philosophy in accordance with this doctrine, following Avicenna and the older tradition of Platonist commentary on Aristotle. But many later scholastics, even within the school of Thomas, felt unable to prove that the soul was separable from the body . . . Aristotelian philosophy began with sensible things, and encountered the soul only at the end of its journey, as the last and most obscure question of physics.[6]

In his ground-breaking book *Beyond Natural Selection* (1993), Robert Wesson hazarded the observation that "the hope of biologists theoretically to base their discipline on physics, the model science, is delusive . . . Living beings operate on a very different level from atoms, and evolution is not a mechanical but a historical process."[7] This is not quite accurate. The opposite of "mechanical" is not "historical" but 'organic', and I believe that the life of mammals has monopolized theoretical thought as a result of the needs of evolutionary and historical accounting. If the significances of insect life and organization enter the equation, a very different result is possible.

The fact is that here in the insect world is the life science equivalent of modern physics' quantum universe. Metamorphosis occurs because the larval form, the pupal form and the adult form have evolved independently of each other to fit different environments. The process has arisen independently perhaps 8 times. Insects such as Lepidoptera and Hymenoptera undergo this complete four stage transformation known as holometabolic. These moths, bees, flies, beetles first appeared in the Permian (220-190 m.) Some other insects—the so-called hemimetabolic, grasshoppers, cockroaches, dragonflies and true bugs—merely go through three steps, reaching maturity by a series of molts omitting the pupa or chrysalis form. This is not merely an issue of the microcosm in relation to the macrocosm. It is rather a matter of organic changes that are common among the insects that have only a marginal echo in the life of a few mammals, such as frogs. Metamorphosis defines the core of quantum biology, a unique and multifaceted process that has implications throughout human culture, both directly and indirectly. The microcosmic transformations are disguised in the mature creature. Insects can destroy their former selves and build a new stage of growth. Larval tissues perish and adult ones emerge from imaginal disks that have been present but undeveloped during the caterpillar's life. It is a sleight of nature, an unusual evolutionary mechanism, and it is built into the structure of our languages.

When Descartes came out of the shadows where he had been incubating his theories, he sensed an insect parallel with his own intellectual development. In ancient Greece the imago signifies the fourfold cluster of butterfly and soul and mind and breath of life in one, ψυχή. And there was a belief that the soul resided in the eye in the form of a homunculus. This is related to the ambiguity of κορή, both a girl and pupil of the eye, and by implication also pupa and pupil-student. Hence when the soul departs from the eye at death, the chrysalis has burst its integument. Ancient Egyptian funerary rites enact this as a central conception, since mummification of the corpse represents and images the chrysalis. Though Aristotle includes a number of groups no longer considered insects—worms, spiders, scorpions, myriapods—for him ψυχή signifies butterfly, and as already mentioned, this is the only Greek word for the butterfly as such, though the word "phalein" was already in use for Lepidoptera in general. Jan Bremmer argues that this would indicate "the meaning will be older than its first occurrence suggests."[8] Aristotle is the first author actually to employ the word 'entomon,' entomology's root.

The Greeks in their pride—or hubris—believed that only a knowledge of Greek would allow an individual to understand his or her perceptions. That is why they classed all non-Greek speakers as "barbarians." Number 16 of the Fragments of Heraclitus makes this clear: "Eyes and ears are poor witnesses for men if their souls (psychai) do not understand the language (literally, 'if they have barbarian souls').[9] And the cosmos, or its Logos and organizing principle, is articulated in the Greek language. The world order speaks to men as a kind of language they must learn to comprehend. Just as the meaning of what is said is actually 'given' in the sounds which the foreigner hears, but cannot understand, so the direct experience of the nature of things will be like the babbling of an unknown tongue for the soul that does not know how to listen. This is apparently the first time in extant literature that the word psyche 'soul' is used for the power of rational thought. Heraclitus is the first to have had something specific to say about psychology of man. For Pythagoras the soul is merely on loan to the individual. Moreover as Erin O'Connell points out, Heraclitus along with other Presocratics "presumes a universal consistency between the logos of the elemental processes of the physical world and the logos of the human cognitive processes."[10]

Could the helix of evolution from ovum through larva then pupa to imago not be what Kant had seen as the "schema" within humans that unifies understanding and the objective world? As he put it in Critique of Pure Reason, "This schematism of our understanding, with respect to its appearance and its mere form, is a hidden art in the depths of the human soul, whose true workings we shall hardly coax from nature or expose unconcealed to view."[11] However as in the relation between classical and quantum physics, the 'coaxing' of the subterranean world into the open is no straightforward affair. Who, after all, would imagine from the visual evidence that humans have so much in common with insects?

Hermann von Helmholtz, an early user of the concept and word quantum in a scientific context, supplemented by Henri Poincaré identified a quaternal pattern of scientific progress as saturation, incubation, illumination and verification, a more general example of the ova-larva-pupa-imago process. And as we shall see, in the sphere of human culture the development of the sonata and symphony follow a parallel course, but one that is more instinctive, an oblique enactment of individual growth. It is ironical that post-Kantian German philosophers from Wilhelm Schlegel to Hegel imagined they were entirely surpassing the original master. It is, though, only in the past 30 years that the full evidence from research into the unity of differing species has been discovered. As Sean B. Carroll explains it: "the genetic revolution in development biology and a series of discoveries that began to unfold in 1984, when evolutionary biologists were again confronted with molecular data that did not fit prior expectations—namely that the disparate body forms and structures of long diverged members of the animal kingdom were governed by very similar sets of genes."[12] In other words mammals may not look like insects, but they are internally identically formed.

So this is a tale of inner parallels between phyla, by definition not apparent to the naked eye. Motoo Kimura writes that "the pattern of molecular evolution is quite different from that of phenotypic evolution," from the growth of appearances.[13] Giraffes and frogs hardly resemble each other, yet at some points of development the embryos do share common features. Likewise, apes together with insects share specific groups of genes that prescribe and regulate central aspects of the body forms (phenotypes) of all animals: "animal body plans emerged using novel signaling and regulatory genes that arose at the inception of multicellular animal life, and . . . once established, the gene-expression patterns underlying the specification of the different body plans have remained fairly invariant."[14] These determining genes were first identified in insects, primarily through genetic and molecular biological research on the long-suffering fruit fly—Drosophila melanogaster. For example the eyes of the flies, mice and humans are encoded by common ancestral (that is, homologous) genes, which suggests that "eye morphogenesis is under similar genetic control in both vertebrates and insects, in spite of the large differences in eye morphology and mode of development." As a result "the traditional view that the vertebrate eye and the compound eye of insects evolved independently has to be reconsidered."[15] And often it is impossible to distinguish insect from ape and from human merely on the evidence of the embryos of the different species. A billion years of evolution have remarkably left the biochemical properties of the determining proteins and their modes of interaction unchanged. This goes back to the earliest of cells with nuclei eukaryotes, and extends backwards to before the Cambrian explosion when diversity expanded some 545 to 490 million years ago.

The great atomic physicist, Wolfgang Pauli, remarked following Niels Bohr that every new philosophy is founded on a paradox. Where the intellect is concerned, the double world of insect-mind corresponds to the wave-particle fusion in quantum physics. Electrons can be produced simultaneously as wave and particle, though no experiment can allow the two to be studied at the same time. So in life, the doublet of insect metamorphosis and mind processes can only be artificially separated, as indeed it is in our Universities and in art to the severe detriment of nature. So Damien Hirst slaughtered 9,000 butterflies for one of his recent works. The particle equates with the living metamorphic creature such as butterfly, while the wave represents its transference into consciousness. After all, we do use the phrase 'brain wave'. They constitute a fused duality. The problem from the standpoint of formal and academic education is that this is a process of simultaneity, whereas academic education tends to encourage a procedure of division and specialization. Certain 'animal' structures inhere in the embryonic mind and are there in the ovum. The clues lie in strings of words conceived during the very birth of language, and apprehended before cultures had ossified into the Greek and Sanskrit mainstream, a type of molten big bang in the language world. It is not, pace Chomsky, that the embryos are forming for sentence structures. Indeed both John Searle and Wittgenstein also inherit from Frege the idea that the fundamental unit of meaning is not in the word but the sentence. So for Searle, the word only has meaning in the context of a sentence. And for Wittgenstein, language is simply a tool, what I call a "common currency," a means to social communication. As Bryan Magee splendidly summarizes the Wittgenstinian approach; there is no essence of language, only social meaning, and we cannot start with private elements and work outwards.

Darwin only revealed the descent from the great apes. But as A.R. Wallace his co-founder of the theory of evolution saw, the intellectual and spiritual evolution of humanity follows a separate line of growth. So we dangle between ape and insect. It was in the Devonian period, some 370 million years ago, that amphibians came onto land heralding the start of metamorphosis. Their progenitors had been the rhipidistians, a genus of fleshy-finned fish. Like today's amphibians, their eggs and tadpoles developed in water but the adults could move on land. They came onto land in the same epoch as early ametabolous insects such as silverfish and bristletails, that is those undergoing very little change in each notch of growth from instar to instar. Insects with complete metamorphosis (holometabolic) diverged from a common arthropod ancestor some 300 million years ago during the Permian. The amphibians remain as makeshift creatures, at home neither in water nor on land. Their metamorphic stages are slow and sometimes regressive. And many marine invertebrates revert to a larval character as late adults, so that sedentary lives are often the culmination.

Bees and moths would appear during the Cretaceous period some 140 million years ago. They seem to have needed several historical attempts before metamorphosis became established. Many modern Lepidoptera were present by the early Tertiary 60-70 million years ago, and by 40 million years ago, all major butterfly families were present. As James W. Truman and Lynn M. Riddiford discovered in 1999, the process of transformation is fired by the way a group of insect hormones, juvenile hormones and ecdysteroids interact during embryonic, larval and pupal stages. These juvenile hormones eventually disappear to become imaginal discs which program the adult.[16] Natural selection favors this sharp differentiation between caterpillar and adult. Vincent Wigglesworth explains: "the sort of body that will be best suited for chewing leaves or burrowing in the carcasses of animals will be very different from the sort of body required for flitting from flower to flower to seek a mate."[17] So the body grub specializes in eating, the imago in mobility, while the pupa is the powerhouse of the transformation. The division of labor, and of form, promotes the survival of the organism. Two different modes of maturation in insects exist. One where the larva or nymph phase passes over to the finished insect without an intervening stage of a pupa, and the other where the full four-set metamorphosis has to be undergone. This dichotomy has a profound significance throughout nature.

George Wald has observed: "a free-swimming echinoderm or unchordate larva, specialized almost wholly for motility and hence dispersal, metamorphoses into a sedentary or sessile adult, specialized for feeding and reproduction [first on plankton and then on the benthos, bottom of the sea]. The winged insects reverse this order: the sedentary larvae, specialized for feeding and growth, metamorphose into highly motile, winged forms, specialized for coupling and reproduction."[18] So holometabolic insects, that is those that pass from egg to caterpillar to pupa to imago, prepare for flight through a full-scale growth. They are the only taxa that pass through a transformation taking them from terrestrial to aerial life. And the angiosperms, from their first flowering plants such as magnolia, follow their evolution and transform via spores to provide the food for the larvae and the nectar for the imagoes, the perfected insects. As Alexandr P. Rasnitsyn's History of Insects records, "Arborescent plants appear in the Upper Devonian, and as Carboniferous insects increasingly dwelled in them to feed, gliding probably became so adaptive for escape and dispersal that flapping wings and powered flight evolved rather suddenly."[19] This dual process of evolution is approximately similar to Bohr's idea of complementarity, where the spectral lines of the structure of atoms revealed through a spectrograph has its equivalent in parallel doublets. The flowering plants are thought to have originated in water and then taken to land via wind-pollination.

This is a fourfold process and why this is such a crux will become clear as we progress. As we shall see this structure in nature has been appropriated by some of the greatest poets and composers. Quite beyond merely descriptive or passionate expression, here the writer or composer threads into the natural world at his and her heart. The integration with that structure meets its point of fusion in the dialectic of microcosm (the poems, sonata) and macrocosm (the dimensions of the universe). A cluster of words expresses the unique outlook of the Greeks. Unlike their predecessors, the Babylonians and Egyptians, the Greeks lived for the present. It was not until Aristotle that the word ὥρα (hora) was adopted to signify "hour." The Greek experience was entirely embodied in the moment; past and future were of no account. While the world of the Pharaohs contested mortality and the present, the Greeks embraced it which is why they could linguistically release the soul as 'butterfly' which fed into Christianity, while the Egyptians rested in the phase of pupation, or mummification.

The so-called Modern Synthesis of the 1930s and 1940s partially updating Darwin and pioneered by Ernst Mayr had, from our standpoint, the paradoxical effect of compounding the insect-ape-mammal succession while emphasizing the bases of evidence that ultimately can lead to an appreciation of the insect connection to humans. So under the evolutionary synthesis: "gradual evolution can be explained in terms of small genetic changes ('mutations') and recombination, and the ordering of this genetic variation by natural selection; and the evolutionary phenomena, particularly macroevolutionary processes and speciation, can be explained in a manner that is consistent with the known genetic mechanisms."[20] However the crucial fields of embryology and developmental genetics played little role in this Synthesis for decades after its definition. Mayr had spelt out the philosophical underpinning: Darwinism has a well-defined philosophical basis, an understanding of which is a prerequisite for the understanding of the evolutionary process. It has long been a puzzle for the historian of biology why the key to the solution of evolution was found in England rather than on the European continent. No other country in the world had such a shining galaxy of famous biologists in the middle of the last century [i.e. 19th] as . . . Germany . . . and yet the solution to the problem of evolution was found by two English amateurs, Darwin and Wallace, neither of whom had had thorough zoological training. How can one explain this? My answer is that philosophical thinking on the continent was dominated at the time by essentialism. This philosophy, as was shown by Reiser, is quite incompatible with the assumption of gradual evolution.[21]

But Mayr has failed to appreciate as other than fragmentary the philosophical and evolutionary insights of Schopenhauer. The full unified coherence of modern knowledge can only be established by accepting the remarkable advances in knowledge prepared by this thinker.

CHAPTER 4

SCHOPENHAUER'S LOST EQUATION

Throughout his intellectual growth, Schopenhauer sought out the structures in nature that corresponded to the mental processes occasioned by his experience of music. This was no dilettantish fad or pursuit of elegance. It was a profound quest for a comprehensive philosophy of nature. For Schopenhauer was aware that he had struck upon the greatest intellectual mystery—what are the pivotal dynamics of the highest of the arts, music? One of the reasons Schopenhauer has not become central to discussions in the English-speaking world is that his reputation has never recovered from the years between 1930 and 1959 when there were only 3 books and no articles in English on the philosopher. And for many, any introduction to Schopenhauer comes via Thomas Hardy who found his pessimism especially congenial.

Here we are speaking of the music of such as Beethoven, Mozart and Haydn. (Schopenhauer specifically praised Beethoven's symphonies). Although at a later date Nietzsche and Wittgenstein will put music at the core of their philosophies, neither grasped the overwhelming significance of classical music's fourfold forms. These forms shape intimate expressions of the will, which is to say of something fundamental to both human and organic nature. Indeed Schopenhauer's philosophy traces, in one great sweep, the quaternities of sonata and symphonic structure. Thomas Mann described his masterpiece, The World as Will and Representation (hereafter WWR), as a symphony in four movements, with the four segments embodied in Books 1, 2, 3 and 4. Indeed Schopenhauer's early work, The Fourfold Root of Sufficient Reason follows the one-dimensional quaternity of medieval thinking inherited from Aristotle, and which he is to explode in The World as Will and Representation. In The Fourfold Root, he catalogues Aristotle's doctrine of the four causes (efficient, material, formal, final) and insists on the difference between Aristotle's cause, and modern causation defined as the relation between events and aspects of life. So one critic has reproached the philosopher, "while interesting, many of his analogies between music and nature are little more than myth or lore."[1] Leibniz had written that "music is an occult practice of arithmetic in which the spirit is unaware that it is counting." Schopenhauer took this to a higher level and saw music as "an unconscious exercise in metaphysics, in which the mind does not know it is philosophizing (WWR 1: 261)." The task is then to ground the metaphysics in organic nature. It may seem an act of hubris to propose to complete Schopenhauer's philosophy. Yet it is now recognized that the most creative and positive act of commentary involves building upon the groundwork of the giants of Western thought and bringing to fruition what still often remain incomplete insights determined by limits in scientific knowledge of earlier times.

Schopenhauer points out that music is recognized as a 'language,' so it must be representational, connected to reality as a copy (Abbild). But since it is not a copy of the world of objects as empirical representations, it must necessarily be of the Will (WWR1:257). "Will" can be read a number of ways, and I shall be reading it as the Other, by which I intend a distinct evolutionary line, and involving humans in forms of intellectual and artistic development that paradoxically link our species to the insects. Schopenhauer was fully aware of the lineage, long before the scientific proof was available:

If we descend through the series of grades of animals, we see the intellect becoming weaker and weaker and more and more imperfect; but we certainly do not observe a corresponding degradation of the will. On the contrary, the will everywhere retains its identical nature, and shows itself as a great attachment to life, care for the individual and for the species, egoism and lack of consideration for others, together with the emotions springing therefrom. Even in the smallest insect the will is present complete and entire; it wills what it wills as decidedly and completely as does man. (WWR2: 206).

Indeed I shall be arguing that what we call 'instinct' often expresses the structures of entomology hammering on the human mental carapace, the cranium.

Schopenhauer writes that if we could succeed in giving a perfectly accurate and complete explanation of music which goes into detail, and thus a detailed rehearsal in concepts of what music expresses, this would also be at the same time an adequate rehearsal and explanation of the world in concepts, or one wholly corresponding thereto, and hence the true philosophy. (WWR1:264). Michael Tanner writes à propos Schopenhauer's approach to music (and defines the issue I am addressing):

For music does, at its finest, for example in Bach's 48 Preludes and Fugues or Beethoven's late string quartets—Schopenhauer is oddly undiscriminating about it—seem to express something that is deeper than the other arts. Whether we can explain this other than by elaborating a full-scale metaphysic into which it fits is a moot point.[2]

And it is the search for this comprehensive metaphysic that is the journey through my book, during which we shall find that the quest opens up evolutionary issues that are generally (unconsciously) avoided or glossed over, when in fact they are the key issues of our day.

John Barrow has observed that "unlike us [the Pythagoreans] didn't think that numbers were just attributes of things. They thought that everything was number. Numbers had intrinsic meanings. They were not just relationships between things."[3] And Schopenhauer wrote that Pythagoras "implanted in our mind the quaternary number, the source and root of eternally flowing creation."[4] Plutarch in Isis et Osiris 75, calls it "Κόσμος, ουρανός, πάν," cosmos, heaven, everything. But Bryan Magee has this excellent remark linking Pythagoras, Schopenhauer and the tetrad:

The fundamental harmonic intervals permeate independently, and always have permeated, the material environment within which man has come into existence, and out of which he is formed (and, among other things, in response to which the biological mechanisms of hearing were evolved). What all this indicates, I think, is that some of our structures of response involving music are programmed into us at much earlier and 'lower' evolutionary levels than anything to do with language—levels which are by countless ages pre-human. And it seems obvious that this fact has a connection with our feeling that music goes deeper than words . . . It relates music at a very deep level with the emergence of man—and hence of consciousness—out of inorganic nature . . .[5]

This musical rooting will then have influenced the emergence of key words as language grew. The World as Will and Representation locates the uniqueness of music in its capacity to replicate the sense of willing. Elsewhere Mallarmé was to value music not for its euphonic elements, but for its structure. Putting these two perceptions together suggests that the human will is predicated on a purposive search for symmetry, that in responding to music the human brain is not seeking out the superficial elements of sound, though it may be these that lead the listener into the labyrinth. When threaded into music, the will intimates some crucial element of human life itself. Music in the form of song associated with work rhythms preceded the literary arts of lyric and tragic verse as Nietzsche suggests in The Birth of Tragedy from the Spirit of Music, so that it appeals to some primary life-impulse. In becoming engrossed in music until a sense of everyday time is lost, we enter the heartland of what I have defined in another book as the Cosmic Cultural Faculty,[6] a faculty which links us purely and directly, if intuitively, to the patterns of the cosmos:

Accordingly, the whole nature of the world, both as microcosm and macrocosm, may certainly be expressed as mere numerical relations and thus to a certain extent be reduced thereto. In this sense, Pythagoras had been right in placing the true nature of things in numbers. But what are these numbers? Relations of succession whose possibility rests on time.[7]

Although there have been various attempts to lump this experience in with the main demands of natural selection, and thus discover a utilitarian and social function, this is merely in line with the routine sociological thinking that has swamped the more subtle processes of thought which in turn has ripped through non-academic civil society.

Steven Pinker goes so far as to reduce the issue to the question: "if music confers no survival advantage, where does it come from and why does it work? I suspect that music is auditory cheesecake, an exquisite confection crafted to tackle the sensitive spots of at least six of our mental faculties."[8] This is too simple and arises from his inability to estimate the impact of the shape of music, its patterning and appeal to human deep structures. So he continues: "music communicates nothing but formless emotion."[9] But as one leading researcher into the musical brain has vividly put it: "When music causes one of these 'skin orgasms', the self-reward mechanisms of the limbic system—the brain's emotional core—are active, as is the case when experiencing sexual arousal, eating or taking cocaine."[10] Actually music expresses four structures; melody, rhythm together with meter, timbre alongside tone, and volume plus dynamic progression. The brain works through a considerable series of processes in order to organize this variegated set of events, and the more experienced the listener or practitioner, the faster these are absorbed. There are specific neural circuits that filter the music. Eckhart O. Altenmüller describes the system:

After sound is registered in the ear, the auditory nerve transmits the data to the brain stem. There the information passes through at least four switching stations . . .[after which] the thalamus—a structure in the brain that is often referred to as the gateway to the cerebral cortex—either directs information on to the cortex or suppresses it . . . . Early stages of music perception, such as pitch (a note's frequency) and volume, occur in the primary and secondary auditory cortices in both hemispheres. The secondary auditory areas, which lie in a half-circle formation around the primary auditory cortices, process more complex music patterns of harmony, melody and rhythm (the durations of a series of notes). Adjoining tertiary auditory areas are thought to integrate these patterns into an overall perception of music.[11]

Schopenhauer wrote that "metaphysics is impossible as being the science of that which lies beyond nature, that is, beyond the possibility of experience."[12] Nonetheless experience itself is ambiguous. Insect metamorphoses seem to be remote from human experience. Even in artistic works, they seem alien except for works such as Kafka's The Metamorphosis, Hardy's The Return of the Native, or David Cronenberg's film 'The Fly.' But if we look at classical music, matters are not so simple. Take the usual structures of symphonies and quartets. A symphony is a sonata for orchestra with, normally, four movements. In the first movement, themes are stated (the egg, or by some recent interpretations, the proto-larva); the second movement proceeds slowly like a caterpillar; the third such as the scherzo of the Eroica tends to be febrile and anticipatory like a shimmering chrysalis trembling with incipient finality; while the fourth usually represents a summation, which as Berlioz analyzed in relation to Beethoven's composition, leads from tension to release, from compulsion to liberation, from the tragic to the joyous. (There are exceptions such as the somber last movement of Brahms's Fourth Symphony.) Undercurrents of profound sadness often underlie the most capricious evolutions of a Beethoven scherzo, and I would interpret this as the process of the loss of imaginal buds which accompanies the emergence of the pupa and later the imago, but nonetheless a loss of youthfulness and an intimation of the end. As listeners or performers, we travel through these stages as the musical form unfolds.

The classical style constructed itself on the four-measure phrase once it had broken from the flowing continuity of the Baroque. The great Cantatas had decorated their music with vocal and instrumental colour like the painted church images. As Spengler puts it: "Music frees itself from the bodiliness inherent in the human voice and becomes absolute. The theme is no longer an image but a pregnant function, existent only in and by its own evolution, for the fugal style as Bach practised it can only be regarded as a ceaseless process of differentiation and integration."[13] So music became more abstract and independent of the Church in the late eighteenth and nineteenth centuries. Charles Rosen estimates that it was about 1820 that the four-measure unit gained pre-eminence in rhythmic structure,[14] but the four-movement symphony had already become dominant even by 1780, although three-movement ones continued to be composed. In other words, the very fabric of the great period of classical achievement is both an anthem to quaternity, and a reverberation of the insect connection. Robert Simpson catches something of this contradiction when he writes that the last movement of Beethoven's Ninth symphony has the composer saying in effect "the visions of the first three movements are such as to reduce man to the apparent size of a microbe; but a man conceived them, so let us all rejoice in our potentialities."[15] The great works of Haydn, Mozart and Beethoven, and those who followed them, are metaphysical to our consciousness, yet have a very real basis in the natural world. Our experience of them is at one and the same time seemingly immediate, and yet ghostly and phantom-like.

More than this, though, there are meta-musical patterns also. Brahms wrote 4 symphonies, as did Schumann, and his final works include the 4 Piano Pieces and 4 Serious Songs. So also of course Richard Strauss's 4 Last Songs. Brahms's Second Piano Concerto does not follow the usual 3-fold concerto shape, but rises to the symphonic four movements. He originally intended his Violin Concerto to have four movements, while his first Piano Concerto was conceived as a symphony based on Beethoven's Choral Symphony. All Brahms's symphonies have a four-movement structure, and all of their first movements follow a sonata shape. Their totality constitutes a hymn to the Tetradic, with their particular thematic continuity and integration through the four sections. They constitute an abstract of the nature of humanity. Schubert also, who so strove to emulate Beethoven, stretched out towards a series of fourfold works, his compositions being "the product of my mind and spring from my sorrow; only those that are born of grief give the greatest delight to the outside world."[16] Three crucial works though are left unfinished: the Quartettsatz in C major, and the E major and 8th B minor Symphonies. As Wilfrid Mellers explains, Schubert "finally solved his most difficult technical and imaginative problem. He had resolved drama into song; and in the andante of the B minor had followed this resolution with the bliss of Eden. He could not rest permanently in a recovered Eden; but at this stage in his career he could not see how he could continue without descent or bathos. He found an answer only in the last three quartets and the C major Quintet, composed during the last four years of his life."[17]

In Blakean terms, Schubert hovered here around the threefold gates of Beulah, while his last four sonatas are the apogee of his writing for piano, perhaps his natural medium with the human voice. Later, Bruckner would build the bar cells of his symphonies using the quaternities of the Mass as his basis. So the final fourth movement of his 5th Symphony unifies the themes of earlier sections in "a double fugue which also embraces elements of sonata style," and which in turn echoes the Agnus Dei of his F minor Mass. Schumann though sensed that the first part of the nineteenth century was the apotheosis of the sonata-based symphony, and wrote in 1835 that he "almost feared that the term 'symphony' might soon become a thing of the past," while by 1839 he feared that "isolated beautiful examples of [the sonata form] will certainly still be written now and then—and have been written already—but it seems that this form has run its life course."[18] (Interestingly these dates approximate to the epoch of the last indisputably great English poetry running from Blake to Shelley and Coleridge, after which Tennyson began to absorb the new sciences.) Although Michael Tippett kicked against the historical archetype of the classical symphony in favor of the notional archetype which permits endless variation, even he ended up writing four symphonies.

Schopenhauer's description of music makes it clear that for him Leibniz's arithmetic of music is only the husk of the issue, a clue to profounder reverberations. Because music appeals so forcefully to humanity's innermost being through a universal language, it goes beyond even the world of perception. And similarly it transcends Leibniz's unconscious exercise in arithmetic, because while that is fine as far as it goes, it only touches the exteriority of the matter and only provides the satisfaction of a sum that comes out right. In fact music appeals to and expresses "the deepest recesses of our nature (WWR1:256). Schopenhauer attributes to music a key role in what was after his death to be defined as natural selection. Music is seen as crucial to the will to survive of the individual since it is a "copy of the will itself, the objectivity of which are the Ideas. For this reason the effect of music is so very much more powerful and penetrating than is that of the other arts, for these others speak only of the shadow, but music of the essence (WWR1:257)." Schopenhauer is offering a purely abstract description which yet makes precise knowledge of the elusive 'thing-in-itself' that Kant could not admit.

Absorbing oneself in a symphony allows the individual to see the whole of her or his life passing along, for "music differs from all the other arts by the fact that it is not a copy of the phenomenon, or, more exactly, of the will's adequate objectivity, but is directly a copy of the will itself, and therefore expresses the metaphysical to everything physical in the world, the thing-in-itself to every phenomenon (WWR1:262)." Music is a familiar yet remote paradise which is perfectly accessible to us, and yet, as Schopenhauer sees it, fundamentally different from our true nature and our environment. What it expresses are "the innermost stirrings of our will, that is to say of our true nature, and since we are not usually aware of these aspects of the essential self, music seems strange and remote."[19]

The strangeness and remoteness of music though arises because it shadows the world of the insects in their metamorphic evolutions, and that unbeknown to us it transports us to that world which paradoxically is also our own internal world of self-evolution. This has a cosmological element as the experience echoes in a ghostly manner contemporary humanity's relationship to the fundamentals of relativity. The listener or practitioner of music is, mentally speaking, in motion as in Aloys Wenzl's description of the twentieth century scientist: "One could say, that the observers of moving systems are like Leibniz's monads and that Leibniz's idea of a pre-established harmony finds an analogy in the theory of relativity. Just as the world is mirrored differently in each monad and yet the sights of all monads are related to each other and translatable into each other, so also does the 'absolute' four-dimensional world-continuum appear in different values of spatial and temporal measurements to every observer imprisoned [as he is] in his own system, yet all sights are transformable into eachother.[20]

Schopenhauer is necessarily working within the postulates of Newton's absolute space and absolute time. In that sense he is shackled by the limits of the intellectual spirit of the age. So when he comes to the more oblique questions, he is drawn into making mystical projections of consciousness: "There is something which lies beyond consciousness but which sometimes breaks into this, like a moon-beam into a clouded and overcast night . . . . It is our essence-in-itself that lies outside time."[21] Music is the clue to the inner nature of the phenomenal world missed by his great progenitor, Kant, and is "so completely and profoundly understood by him in his innermost being as an entirely universal language, whose distinctness surpasses even that of the world of perception itself (WWR 1:256)." It is "as it were the innermost soul of the phenomenon without the body (WWR 2:262)."

Bryan Magee in his book on Schopenhauer argues that

> the reason why words cannot dig down to the same level [as music] lies in their excessive generality . . . . Language cannot make use of concepts which are formed by a process of generalization, and this means that it can never communicate insight into the unique in-itself-ness of anything. Music, however, does. And in doing so it is 'completely and profoundly understood by [man] in his innermost being as an entirely universal language whose distinctiveness surpasses even that of the world of perception itself.' Now since what a philosopher like Schopenhauer is trying to do is to formulate an 'expression of the inner nature of the world in very general concepts', it follows that the composer is already doing in concreto what the philosopher is attempting to do in abstractio. Therefore if, per impossibile, we could succeed in giving a 'perfectly accurate and complete explanation of music which goes into detail, and thus a detailed rehearsal in concepts of what music expresses, this would also be at the same time an adequate rehearsal and explanation of the world in concepts, or one wholly corresponding thereto, and hence the true philosophy . . . . The composer reveals the innermost nature of the world, and expresses the profoundest wisdom, in a language that his reasoning faculty does not understand.' His is therefore the purest, the most undiluted form of genius of all, because it is the least contaminated by conceptual thought or conscious intention.[22]

It is the per impossibile that is the challenge. This is the molten lava of the many unanswered questions and crucial links between science and linguistics; in other words what knowledge of the noumenon depends on, and where it can be found. Get to grips with this, and there will be an extension to the theory of evolution, distant though for the moment the professional scientific world is from admitting this as an issue. Whereas Relativity theory places all time on one curve so that past, present and future are contemporaneous, quantum physics makes time entirely uncertain, to the degree that its next move is unknown in advance—it is wave and particle. But neither of these can explain the human (subjective) sense of the flow of time. It has been suggested that musical composition, with its evocation of the passing of time which is yet brought to closure by the structure may offer some clues here. I suggest in my The Metaphysics of Insect Life that the sense of time elapsing as experienced in listening to (the usual) 4 movements of a symphony or quartet is also a passing through 4 major stages of intellectual and spiritual life corresponding to the 4 stages undergone by those insects that undergo metamorphosis. In other words, playing or listening to the key works of Beethoven, Mozart, Haydn, and perhaps especially Brahms, re-create our sense of time passing, of growth within an instinctively structured setting. Indeed these works are an experiential clue to the inner (genetic?) meaning behind language, quite beyond the structural grammars essayed in Terrence Deacon's epic The Symbolic Species.

Clusters of words also discussed both in my *Metaphysics* and *The Poem and the Insect* reach deep back into the formation of language, and represent both a response to surrounding nature, and some inner hinge that connects us not only to other primates, but from a more evolved cultural level, paradoxically, to the insect.

CHAPTER 5

ANDRE MALRAUX: LITERARY QUANTUM CREATOR

Insofar as the label "modern Renaissance Man" retains currency value, then André Malraux is the most deserving candidate, and as a politician reduces Thatcher, Blair, Obama, and the Bushes to philistine mediocrity. Novelist, politician, organizer of the Republican air force during the Spanish Civil War, philosopher of art—he placed metamorphosis at the center of both political and literary strategies. The centrality of the example of the International Brigades during the Civil War when individuals from all over the world volunteered to fight Franco and the Fascist forces from Germany and Italy remained—and indeed remains—an inspiration. In Malraux's novel of the war, there is as much a battle against the human condition being fought out by the Republicans as one against the enemy. As René Girard put it: "the enemy is only one element of these cosmic forces that crush man and against which the heroes have thrown down a challenge."[1] So the hope at the end of *L'Espoir* (Days of Hope) is that "men will be voluntarily metamorphosed, just as Manuel allows himself to be."[2] From his early fantastical and rather jejeune writings, Malraux appreciated and celebrated the mystery and power of this natural process, something which he argues in his mature work of *The Voices of Silence* and *The Walnut Trees of Altenburg* determine the rise and fall of civilizations. He embodies the quantum mind, wave-particle dualism, able to move from fluent theorizing to specific analysis seamlessly, not least in his works on art. In another context Frohock unwittingly describes his remarkable mind "that leaped from idea to idea so abruptly that his reader inherited the task of supplying the transitions, in particular the causal ones."[3]

A startling statement in The Voices goes so far as to draw a parallel between the life of the insects and humans:

Insects' tools are the limbs with which they are equipped from birth and which they cannot change; but genius puts forth unseen hands which, throughout the artist's working life, are ever changing and enable him to extract from forms, both living forms and those immune from death, the makings, often unlooked for, of his metamorphosis.[4]

The miracle of transformation is that, as in the case of Lepidoptera, the early larval imaginal cells or discs (in a writer's case, his or her juvenilia) contain the 'seeds' of the later mature writer. As the Dictionnaire Alphabétique puts it, it is "a change of form, of nature or structure so extensive that the being or creature that undergoes it is no longer recognizable." Möllberg in *The Walnut Trees of Altenburg* declares: "I doubt if there's any communication between the caterpillar and the butterfly. Even between the Hindu who believes in the absolute and the transmigration of souls, and the Westerner who believes in the fatherland and in death, any communication is artificial"[5] As late as 1974, two years before he died, he wrote "I am attempting to create a life of someone at a moment in history when we are becoming aware of Metamorphosis as Universal Law."[6] He points out that fragile as they look, butterflies have survived on the planet for 260 million years spanning the Equator to the Himalayas[7]. This longevity of the species belies the ephemerality of their lifespan, and for Malraux it takes on the aspect of the sacred.

Gisors in The *Human Condition* asks Malraux's primary question: «Que faire d'une âme, s'il n'y a ni Dieu ni Christ?» (What can we make of a soul if there is neither God nor Christ),[8] and goes on to remark that "the ideal of a god is to become a man while knowing he can keep his power; the dream of a man, to become a god without losing his personality." Jean Carduner interprets this in Malrucian terms as "every man dreams of a reversible metamorphosis, becoming a butterfly while retaining the possibility of returning to the caterpillar state."[9] I take this to mean that individuals who have progressed through the stages of the fourfold evolution may wish to return to a lower stage so they can keep in touch with those only beginning the ascending journey.

However Malraux also sees a negative quality in the insect. Death in his early work *Paper Moons* is "a huge insect," while in The Conquerors, "the deeper the narrator penetrates into the Oriental world, the more often he encounters the sinister, emotionless insect." When in The Royal Way, Perken is about to deal with the chief of the Moïs with their weapons of lances and cross-bows he is confronted by "antennae," "mantis arms" and "mandibles," and found he "depended totally on this being with his larval thoughts."[10] In his final novel, The Walnut Trees of Altenburg he "finally relinquishes the pejorative threat of the insect analogy."[11] "And in his 1932 article, "Jeune Chine," Malraux defines:

> the true life which animates the metempsychosis of the enormous insect of China is the battle of various ideas against deaf forces," an overcrowded society which cannot but deny individuality, filial duties that are supreme and an attachment to the dead to the extent that the dead are the living. The Eastern world follows "a geological rhythm in which man flitted past like a butterfly.[12]

It is enshrined in Byzantine art as "a supreme negation of the ephemeral." According to Malraux, the contemplative unity of East and West died in 1911 when the West brought in modern man living in "le temps du silex" (the age of flint) and wishing to dominate others, while the East continued to submit in order to approximate to God. Nonetheless the West continues to lead the battle against determinism, itself a test of human determination which has been the source of the West's creativity and its great individual works of art. As he puts it, in the war against fatalism "every masterpiece is a human victory over the blind forces of destiny."

But farfelu is the motto for his outlook in both East and West, and he identifies an appearance in the church windows of the West where it represents "the eternity of the ephemeral." His wife, Clara, recalls that he told her that "this word appeared very early in the French language and that one common root linked it to the Italian word farfallo, 'butterfly.'[13] (Surely Malraux knew the poems of Guido Gozzano, the greatest poet of the butterfly.) In an interview, Malraux identified the source of farfelu as Rabelais: "Rabelais used the ancient etymology which implicated the egotistic swollen bombasts, and at the same time drew on the false etymology farfalla—'papillon'—and the mélange gives you the boaster with butterfly wings, or farfelu." Malraux called himself "farfelu numéro un." The importance for Malraux the novelist was that filtering reality through the farfelu, he "was capable of forming an autonomous and strictly imaginary world which existed independent of the restrictions of conventional reality. By means of his creative volonté, the writer fashioned a new kind of artistic reality which rivaled and surpassed the original on which it was based."[14] This is in some way related to Möllberg's question in Les Noyers clearly not endorsed by the author: "is it possible to identify one permanent fundamental idea, valid everywhere throughout history, which underpins the beliefs, the myths, above all the multiplicity of cerebral structures, and on which the notion of man can be based?"[15] This may not be endorsed by Malraux, but it lies behind much of his thinking and action, while many of his compromises with the Communist Party in the late 1930's and later with the Gaullists, are but the spume on the surface of his life. "What Western youth is seeking is a new notion of man,"[16] he writes in his early "D'une jeunesse européenne." Because death renders everything in life vain so far as the West is concerned, the empire of death must be called "The Royal Farfelu." So in the Bhagavad Gita, huge butterflies come and land after combat on dead warriors and sleeping victors. The image of "the shadow-insect" is often used to define those humans who have not outfaced a given destiny imposed upon them, and created one of their own. As Violet Horvath put it: "the only beings are those who create. All others are shadows floating on the earth, strangers to life."[17] In Malraux's progressively darkening vision, destiny becomes increasingly hostile to humans. And it is with death that destiny achieves its defining triumph: "death, then, is the ultimate metamorphosis, transforming what was our life into a permanent, irremediable destiny." All the debates among characters in *L'Espoir* as to the superiority of doing over being, have finally been reduced to the academic.

In his *The Mirror of Limbs* Malraux always associates the butterfly with the supra-logical. And as David Bevan has noted, there is an "implicit association of an affinity for butterflies with a capacity to perceive beyond the purely logical,"[18] and this takes us to the realm of quadrilectics well beyond dialectics. It rivets the processes of mind to the natural external world. The attractiveness of Lepidoptera encourages the insinuation of the insect's processes into human consciousness. In his early writings Malraux had linked the science of the Einstein-Heisenberg era with the crisis in human identity as old certainties disintegrated:

It seems that our civilization tends to create for itself a metaphysic where every fixed point were to be excluded, just as with the conception of matter. The human and the self, one after the other destroyed . . . creating a domain of spirit and sensibility defined by movements, changes, new relations and new births.[19]

He wondered whether a new attempt to understand the nature of the human species would be undertaken when appreciation of metamorphosis came to dominate the theory of evolution.[20] And he quotes De Gaulle in his *Felled Oaks* to provide the motto for the late twentieth century crisis:

"I pick up these books again," says De Gaulle, they have survived and have perhaps shaped men, as successive gardeners have shaped my trees."[21] The absence of 'a shaping man' becomes a late obsession for Malraux, especially in the wake of the events of '68, because he saw that the worship of science despite its method and analysis offered—and continues to offer—"no ordering value."[22]

CHAPTER 6

WOLFGANG PAULI'S BRIDGE BETWEEN ART, PSYCHOLOGY AND SCIENCE

The Nobel atomic physicist, Wolfgang Pauli remarked, as Niels Bohr had before him, that every new philosophy is founded on a paradox. Where the intellect is concerned, the double world of insect-mind corresponds to the wave–particle fusion in quantum physics. Light has the double nature as simultaneous wave and particle, though no experiment can be set up to allow the two to be studied at the same time. So in life, the doublet of insect metamorphosis and mind processes can only be artificially separated. The particle equates with the living metamorphic creature such as butterfly, while the wave represents its transference into consciousness. After all, we do use the phrase 'brain wave'. They constitute a fused duality, and suggest why the philosophy I explore has proved so elusive.

Pauli was awarded the Nobel Prize for his elaboration of the exclusion principle whereby the state of an electron in the shell structure surrounding the atomic nucleus in chemical elements is defined by four numbers rather than the three quantum numbers. "In association with modern dreams, Jung saw movement from three to four as symbolizing a stage of inner development known as the individuation process. Pauli saw his discovery of the exclusion principle in that light."[1] And in alchemy as well as in Jung's psychology, moving from three to four symbolizes a completion, or a movement toward the centre. Although he insisted that "quaternity would not be effective in physics,"[1] Pauli saw that "a quaternity would certainly be assigned to the wholeness made up of physics and psychology insofar as the complementary pair of opposites of physics reappears reflected in the psychic sphere."[2] This involved an expansion of physics, perhaps in conjunction with biology, so that the psychology of consciousness could be integrated. This expansion had implications for philosophy. Bohr had tried to solve the problem creating a mini-solar system as a demonstration. But Pauli realized that in understanding the behaviour of electrons in relation to the atomic nucleus of the chemical elements, it was impossible to visualise the situation. Science had moved into purely abstract territory.

And he defined four domains of natural laws. These are classical physics ruled by causality; the special theory of relativity in which space and time constitute a continuum; the general theory of relativity; and quantum mechanics in which the classical concept of causality breaks down and is replaced by quantum-mechanical complementarity. So here he is arguing that to confine the issue to physics alone is cosmologically an error. As he put it—"This is not a question of any shortcomings of the quantum theory within physics but a shortcoming of physics within life as a whole."[3] Indeed the confidence in nature's completeness engendered by quantum discoveries led Pauli to question Darwin's theory itself. Nature perhaps contained a tendency toward fourfold perfection, not so much a teleological aspect, but an organic structure that insisted on attaining completeness. The possibility that a fourfold structure permeated the physical as well as the psychic dimension was for Pauli more than a curiosity. Coupled with the idea of synchronicity as a creation in time, it eventually led him to question the randomness of natural selection as proposed by Darwin, arguing that the process of evolution was directed toward the goal of completeness.

Pauli did not confine his theorizing to quantum science. Together with Carl Jung, he sought a fusion between quantum processes and human evolution. He went so far as to propose a 'third law' of nature that accounted for nature's drift toward wholeness. Pauli proposed that "a third law of nature was required since:

> evidence showed that evolution was not purely a matter of 'blind chance.' He asserted that random mutations could not account for certain evolutionary developments, and contrary to blind chance, that mutation of genes had probably worked purposefully and meaningfully for the human species.[4]

The core of this was integration with the life sciences: "The future development must, under the influence of the out-flowing current of unconscious content from psychology, involve an expansion of physics, perhaps in conjunction with the life sciences, so that the psychology of the unconscious can be accepted there."[5] This would be in line with Husserl's strictures on physics and its alienation from real life.

One of the major emphases of quantum physics is the duality of wave and particle, a fact stumbled upon by Einstein although he never embraced quantum theory. Complementarity accepts the paradoxical quality of nature as being understood, whereas the concept of not-being as applied to the paradox avoids it. No experiment can be set up to identify both particle and wave since they manifest themselves simultaneously in the atomic physics. This physics Pauli argued: "was related towards an end, a fitness for purpose, and wholeness which we regard as characteristic of life and living things. This raised for him the question: Is it only in the association of physical and psychical processes, and not in other situations as well, that a parallelistic relation exists? And do not these relationships conceptually embrace 'a unity of essence?' This suggested to Pauli a depth of meaning in the nature of matter that extends to all forms of existence."[6] In stressing the spiritualization of matter in Answer to Job, Jung had said nothing about the materialization of spirit, and it was important to Pauli that it be mentioned. What was neglected in Answer to Job was the materialization of the spirit, or the concrete or chthonic dimension of the spirit. Jung could not placate Pauli with some compromise form of words, and the scientist insisted that psyche and matter belonged together.

Pauli goes on to consider the notion of the bridge between the sense perceptions and concepts:

It seems most satisfactory to introduce at this point the postulate of a cosmic order independent of our choice and distinct from the world of phenomena . . . [T]the relation between sense perception and ideas remains predicated upon the fact that both the soul of the perceiver and that which is recognized by perception are subject to an order thought to be objective . . . The process of understanding nature as well as the happiness that man feels in understanding—that is, in the conscious realization of new knowledge—seems to be based on a correspondence, a 'matching' of inner images pre-existent in the human psyche with external objects and their behavior.[7]

So here in my analysis and adaptation of the quantum, particle and wave correspond to the processes of psyche-based thought with which they are fused in the Greek word ψυχή. Once the foundation has been absorbed into the mind, science, language and the arts can begin their long osmosis. Elsewhere Pauli drew our attention to the similarity between the radioactive nucleus and the self. Both undergo transformation and affect their surroundings. He interpreted the stranger in his dreams as someone in need of redemption by transformation. The nucleus at the core of an atom disintegrates, just as radium is metamorphosed into lead as it decays. In parallel, the self changes itself as its former manifestation atrophies. Pauli wrote that

> The four chambers of the heart seem to have a relationship to the quaternity of the mandala. Therefore, to me there is the question whether there are not parallels to the course of the individuation process (with its psychological development of the center) in comparative anatomy of animal lineage or in the embryonic development from lower animals with single blood vessels up to the formation of the heart.[8]

The implication was that, as with individuation, there was an underlying trend, not only psychically but biologically as well, toward expressing the fourfold aspect of Jung's mandala in nature. Indeed we can add that while there are 4 aorta chambers in hot-blooded mammals, there are just 3 in reptiles and the cold-blooded.

At this point Pauli for the first time here expressed a view that broke ranks with Darwin, arguing that the process of evolution was directed toward the goal of completeness. The possibility that a fourfold structure permeated the physical as well as the psychic dimension was for the scientist more than a curiosity. This links with Richard Dawkins' observation in The Selfish Gene that DNA is "written in the A, T, C, G, alphabet of the nucleotides. It is, as though, in every room of a gigantic building, there was a book-case containing the architect's plans for the entire building. The 'book-case' in a cell is called a nucleus." Pauli writes that "the ancient Pythagoreans, with their reverence for the number four, would have taken particular pleasure in the quaternary chemical structure, built up on two pairs of opposites, of a nucleic acid (denoted DNA for short) which is essential for the processes of heredity and reproduction."[9] Pauli's intuitive feeling that:

> the 3-4 problem could be solved only by living the 3 and the 4 simultaneously—in other words, by relating to the dynamic aspect of the self. In his dreams, analyzed in conjunction with Jung, the 'irrational' relationship of the diagonals to the sides of

the square (mathematically speaking) symbolized the premise that the 3-4 way of seeing reality cannot come about by rational thought alone. Von Franz, who years later made a study of 'numbers and time,' referred in her book Number and Time to Pauli's diagonals dream in the context of the 3 and the 4, noting that the 3, or Trinitarian thinking, represents the dynamic process of thought that flows linearly with time and leads to a concrete result, such as a dogma, with no absolute validity; in contrast, the 4, or quaternary attitude, develops 'more modestly based' on archetypal concepts.[10] The relation between the numbers 3 and 4 obeys a similar indeterminism that characterizes quantum physics, and thus overcomes the contradiction between wave and particle.

This also has implications for gender organization, since 3 tends to be the number of female systems, and 4 of male. Pauli writes that "real pairs of opposites, like particle versus wave, or position versus momentum, or energy versus time, exists in physics . . . One number of the pair is never elemental in favor of the other, but both are taken over into a new kind of physical law which expresses properly the complementary character of the contrast." As a result of the mental and personal battles of Wolfgang Pauli, he surmised there is an almost Manichean war in the cosmos and human psyche between the numbers three and four. As he put it: "Modern quantum physics has come closer to the quaternary point of view, which was so violently opposed to the natural science that was germinating in the 17th century." Robert Fludd it was who, railing against Kepler's espousal of the trinity in the early days of science, had exclaimed "you force me to defend the dignity of the quaternity" (cogis me ad defendam dignitatem quaternarii).[11] So Pauli believed that in the interest of maintaining his own equilibrium he must continue to nurture the elements of scientist and mystic within himself as he asserted in a letter to Fierz in 1950: "that I carry 'Kepler' as well as 'Fludd' in myself and that it is a necessity to arrive at a synthesis of this pair of opposites as best I can." He could not commit himself to Kepler entirely because his discovery of the electron's state had depended on his adherence to a fourfold nature and a fourth quantum number.

The author of *Deciphering the Cosmic Number*, Arthur I. Miller, comments: "the archetype of the wholeness of man—depicted with the symbol of fourness, the quaternity—is the emotional dynamic that drives all science."[12] The same archetype drives the great achievements of major literature and music. So in the seventeenth century Kepler "whose heliocentric convictions were based essentially on the fact that he saw the Copernican world system as a symbol of the Trinity." Pauli further writes on Kepler: "The author's critical admiration for the manifestation and the utterances of his 'anima' may well be the reason why he gets bogged down with the triads and never makes it as far as the quaternity."[13] It was a result of Pauli's intuitive adherence to the quaternal that enabled him to conclude that each electron in an atom required four and not three quantum numbers, and that no two electrons in an atom could have the same quantum values. Miller speculates that Pauli had tapped into something beyond science, something touching on one of Bohr's favorite quotations—from Schiller:

> Only fullness leads to clarity
> And truth lies in the abyss.

There was no Dantesque pilgrim to lead the scientists who undertook this journey of discovery which effectively confronted in the abstractions of physics the dissociation of sensibility, as T.S. Eliot termed the seventeenth century crisis of fractured thinking. From his intellectual exchanges with Pauli, Jung was able to draw out of the abyss this dramatic insight: "How could the quaternity arise in the unconscious? There must be something in the psyche adhering to a fourfold world of individual realization." This he saw as the source of "the existence of an archetypal God-image" in the human mind. But Pauli gave it a more materialist and natural spin—

The process of understanding nature as well as the happiness that man feels in understanding—that is, the conscious realization of new knowledge—seems to be based on a correspondence, a 'matching' of inner images pre-existent in the human psyche with external objects and their behavior.[14]

The re-association of the dissociated sensibility—the achievement of true balance in the individual—can only be accomplished through the continuous battle to fuse science and the arts. Pauli, as a result of his dreams, constructed a lexicon of association exemplified in two analogies culled from nuclear science:

"1. The splitting of spectral line by a magnetic field to form a doublet he equated to the splitting of an archetype into its opposites as a step toward a differentiated consciousness.

2. The radioactive nucleus he identified with characteristics of the self. Both are said to undergo transformation, and both radiate into their surroundings. The radioactive nucleus at the center of the atom undergoes disintegration (as, for example, radium is// transformed into lead through various stages of decay). By analogy, the self is understood to be capable of transformation, and it is believed to have an effect on its surroundings (as if by radiation)."

3. Time: "'Certain time' was experienced linearly in terms of past, present, and future, whereas 'uncertain time' connected with a synchronicity, in which a moment seems to have a quality that relates meaningfully and unpredictably to the future."[15] Later in his career Pauli spoke of the "larger spiritual transformations" in a letter exhibited at CERN. For in microphysics the observer is different in an essential way from the "detached observer" of classical physics and compares the effect of an observation of a quantum system (the "reduction of the wave packet") with a transformation (Wandlung) in the alchemist sense. He thought that in physics the remedy for the too complete detachment of the observer may lie in the integration of the subjective, psychic. Indeed, in Science and Western Thought Pauli asks the question:

Shall we be able to realize, on a higher plane, alchemy's old dream of psycho-physical unity, by the creation of a unified conceptual foundation for the scientific comprehension of the physical as well as the psychical?[16]

This quest for a unity of physis and psyche is a recurrent theme in the exchange between Pauli and Jung and is the main concern in Pauli's Background Physics in which he was guided by his dream motifs. It was also a motivation for his Kepler article in which he describes the polemic between the rational Kepler representing the new scientific attitude and the irrational Fludd who defends the old alchemist world view. Kepler, like Newton after him, firmly believed in the Trinity of the Christian God while Fludd got his inspiration from the Pythagorean quaternity which for him was a symbol of the unity of the world. Pauli admits that his sympathy is not only on Kepler's side, remembering that the discovery of the exclusion principle had been possible only after realizing that the electron's state depends on a fourth quantum number. Pauli magnificently defined this continuing dilemma, which is unresolvable except by taking refuge in the particular blindness to one side of the multi-faceted problem which only today's scientific professional bureaucracy can afford. As I have already quoted—and it bears quoting again—Pauli characterizes Fludd's vision of a fully integrated outlook as follows:

> Even though at the cost of consciousness of the quantitative side of nature and its laws, Fludd's 'hieroglyphic' figures do try to preserve the unity of the inner experience of the 'observer (as we should say today) and external processes of nature, and thus a wholeness in its contemplation—a wholeness formerly contained in the idea of the analogy between microcosm and macrocosm but apparently already lacking in Kepler and lost in the world view of classical natural sciences.[17]

> The modern development of quantum physics has raised the possibility of resolving this conflict, and establishing a wholeness of outlook which can incorporate the fields of music and literature as well.

CHAPTER 7

HUSSERL'S HEROIC DEFENCE OF HUMAN NATURE

Husserl's final work, The Crisis of European Sciences (1935), was born of his removal by the new Rektor Martin Heidegger from the University of Freiburg, where he held emeritus status. Heidegger was his former assistant, but had now embraced the Nazi State. As Paulius Ryanto has written, "the majority of University Rektors [in Germany] favoured variations of a folk-philosophy mission from outright Nazi to anti-liberal authoritarianism."[1] By 1929, Husserl had rejected the concept of Heidegger as his disciple; "I came to the conclusion that I can not admit his work within the framework of my phenomenology, and unfortunately that I also must reject it entirely as regards its method, and in the essentials as regard its content."[2] The 'history-element' in the later Husserl writings emerged in the context-horizon of his own awareness of 'the crisis' of European culture, which he had to admit had been brought about by modern science and philosophy. But the shift also came about by Husserl's own late-coming dialogue with history. This dialogue was somehow forced upon him when he had 'to make sense' (Besinnung) of being required to leave Freiburg University as a prominent thinker and in his emeritus status through the actions of the new rector, his former assistant Martin Heidegger, as well of the Nazi State."[3] The majority of University Rektors favored variations of a folk-philosophy mission from outright Nazi to anti-liberal authoritarian.

Heidegger dismissed Locke's distinction between persons and human beings.[4] This is as authoritarian as Rousseau's 'social contract' which fastened citizens in the chains of social conformity. It was this contract that Benjamin Constant had exposed as the sleight of mind in Rousseau's presentation of freedom.: "the conquerors of today are resolved to gaze over the level surface of their empire and to encounter no deviation from

70

uniformity . . . It finds the isolated individual easier to deal with; without effort it crushes him beneath its mighty weight." Husserl diagnosed contemporary science and philosophy as susceptible to skepticism and irrationalism—"merely fact-minded sciences make merely fact-minded people" since science had been reduced to fact-finding.[5] Heidegger radically opposed "all recognition of the individual value of the human spirit, and therefore of the whole of modern philosophy as developed on the basis of Descartes."[6] But according to Husserl, the Renaissance had already opened up a new freedom, choosing its model as the classical humans who were defined by philosophy. The history of philosophy in this interpretation is an historical search for the "meaning of man" (sinn des Menschen). And European philosophy filters out mythology, indeed the type of mythology that Heidegger and the thousand academics were instituting via the Nazi Party.

In his later book The Crisis of European Sciences and Transcendental Phenomenology (1935), Husserl deepened his critique of positivism as well:

Positivism, in a manner of speaking, decapitates philosophy. Even the ancient idea of philosophy, as unified in the indivisible unity of all being, implied a meaningful order of being and thus of problems of being. Accordingly, metaphysics, the science of the ultimate and highest questions, was honoured as the queen of the sciences; its spirit decoded on the ultimate meaning of all knowledge supplied by the other sciences.[7]

Heidegger's hostility to metaphysics was part of "the collapse of the belief in a universal philosophy as the guide for the new man [and] actually represents a collapse of the belief in 'reason,' understood as the ancients opposed episteme to doxa."[8] So this last work of Husserl can be read as a last attempt to reassert against the trend toward irrationalism, "the infinite power of reason and the ideal of a perfect, all-embracing science." A universal and grounded phenomenological science was required "as making possible mankind's development into a personal autonomy and into an all-encompassing autonomy for mankind—the idea which represents the driving force of life for the highest stage of mankind." Herbert Spiegelberg emphasized the starting-point of Husserl's philosophy as being his critique of the British empiricists:

The British empiricists from Locke to Hume were Husserl's introductory readings in philosophy and remained of basic importance to him all through his later development. Often he gave them credit for having developed a first though inadequate type of phenomenology.[9]

For Heidegger the self derives its identity from the way it understands and interacts with its environment, thus anticipating any number of dictatorships of conformity. This is the crux of Heidegger's attack on Husserl, his destruction of his public life and attempted erasure of the influence of his work:

The concept of private experience ('Erlebnis') provides the methodological starting-point for Husserl's investigation of the different kinds of objects that populate our shared, public and objective world and the structures that allow us to understand that world. Heidegger rejects the notion of a private experience, indeed the very notion of Erlebnis, that has its heyday at the end of the nineteenth century and in the early twentieth century. However, Heidegger continues to give central importance to the other German term for experience, Erfahrung. The notion of experience lacks the connotation of private, subjective experience that is characteristic of the notion of Erlebnis.[10]

Heidegger rejects the idea that a human being has an essence, and argues that an individual has no value, except in a community of blood. If Husserl's concept of a human essence seems unduly mystical, then Heidegger's is brutally idealist and mechanical; he "develops his conception of self out of an almost neo-Kantian reconstruction of Aristotle's idea that the soul is the for through which the life of the body is organized."[11] As Heidegger started his lectures he attacks Husserl for "privileging the pure transcendental ego over what Heidegger at this point called the "historical ego" and the "ego of the situation."

In his Cartesian Meditations, Husserl argues that "anyone who intends to become a philosopher must 'once in his life' withdraw into himself and attempt, within himself, to overthrow and build anew all the sciences that, up to then, he has been accepting."[12] Transcendental knowledge is a philosophy of possibility, and this depends upon taking possession of "the absolute ground of pure pre-conceptual experience, which is its own preserve; then, self-active again, it must create original concepts, adequately adjusted to this ground, and so generally utilize for its advance an absolutely transparent method."[13] His take is that the 'nature' of the organism, and in combination with it, the soul, become constituted in interrelation with each other, in unity with each other.[14] The interaction with the environment, pace Heidegger, is secondary and bracketed.

Although Husserl starts from the British empiricists, and although he praises empiricism's respect for things themselves, he pinpoints its weakness as restricting intuition to that sensory experience. This is where Husserl's Phenomenology comes in. The task of phenomenology, as Kolakowski explains it "is not to describe a singular phenomenon but to uncover in it the universally valid and scientifically fruitful essence, or eidos. The eidetic insight, however, is not a procedure of abstraction, but a special kind of direct experience of universals, which reveal themselves to us with irresistible self-evidence."[15] This is especially significant in the light of Max Müller's theory of the historical formation of language roots as described in Chapter 1, which assumes a method based on a similar, if primitive, mode of apprehension.

The seventeenth century Leveller and follower of John Lilburne, Richard Overton, puts the issue for the individual more plainly even than Locke, who wants to recruit the individual for a definition of 'person' who has property in his or her self, and so in the course of time the right to vote:

To every Individuall in nature is given an individual property by nature, not to be invaded or usurped by any: for every one as he is himself, so he hath a selfe propriety, else could he not be himself, and on this no second may presume to deprive any of, without manifest violation and affront to the very principles of nature, and of the rules of equity and justice between man and man . . .

Locke is the philosopher most identified with psychological continuity as the criterion of personal identity. In Locke's classical defense of the forensic concept of a person, the primary bearer of responsibility is consciousness rather than the human being who has that consciousness. The unity of consciousness is what Locke calls a person." Heidegger's failure: "his attempt to discover a ground for language, without which no authentic experience of things or of the world may be attained. The problem is that language cannot be grounded in something other than itself because 'Language languages' just as World worlds."[16] From the period of Being and Time, Heidegger sees in the philosophy of Descartes "the very essence of dualistic, logical representational, scientific, mechanistic metaphysics."

Heidegger's Dasein (Being) has no integral necessity; it is merely cobbled together on the basis of some sort of peasant mysticism. His fourfold (Das Geviert) may have originated in an intuitive sense of the importance of fourfoldness, but it lacks any grounding in culture. This

Das Geviert of earth, sky, mortals and gods is not even a meaningful metaphysics, which of course has been dismissed with his judgment that Nietzsche was "the last metaphysician." Since Heidegger later insists that authenticity is assured by the fourfold is a serious lacuna. Heidegger deconstructs his quadrate by imagining it to be a pouring-out from pitcher which constitutes 4 aspects of Being. Water wells up from the earth, wine from the sun-drenched vine, springs are fed by water from the sky, and mortals quench their thirst. By this time (1950's) he has abandoned ancient Greece as "the true 'origin' of the Western understanding of Being" and now 'dwelling' has become the 'essence' of Being, again some sort of almost peasant ideal. All this is in line with his elevation of violence as the catalyst for change:

Before Being can occur in its primal truth, Being as the will must first be broken, the world must be forced to collapse and the earth must be driven to desolation, and man to mere labor. Only after this decline does the abrupt dwelling of the Origin take place for a long span of time.

But the key elements of language as part of an "inquiry-ab-ovo as Descartes insisted gives a sound technical and historical origin to the fourfold and puts it at the center of philosophical thinking. All Heidegger's strictures aimed at Husserl and defining the world as a place of coping, are little more than hyper-active bluster and bullying. Twenty years after his Being and Time, he admits "the fitting word is still lacking even today." Ironically in the light of his full endorsement of Nazism, "for Heidegger, what arrives in the wake of philosophy's end is above all the demand to requisition the meaning of language—of that which makes us human." The use of the word "requisition" succinctly diagnoses the violent impatience of Heidegger's method. He short-circuits the processes, and substitutes activist coercion—blood and soil—for a true solution to this problem of origins.

CHAPTER 8

NATURAL EVIDENCE FROM SHAKESPEARE

Shakespeare tends to write in segments of 4 plays. So the peak of his early work is the two historical tetralogies. Later there are the 4 great Tragedies where his heroes and heroines meet their fate in 5-Act worlds. These are interwoven with 4 Problem Plays followed by the 4 Last Plays. Keats's perception of Shakespeare's biographic parabola can be seen anew in the light of these paradigms:

A Man's life of any worth is a continual allegory—and very few eyes can see the Mystery of his life—a life like the scriptures figurative—which such people can no more make out than they can the hebrew Bible. Lord Byron cuts a figure—but he is not figurative—Shakespeare led a life of Allegory; his works are the comments on it.[1]

Working from this analogical method, the following reveals itself:

❖ 1. The Egg (the History Tetralogies)—3 Parts of Henry VI, Richard III; and then Richard II, 2 Parts of Henry IV, and Henry V.

❖ 2. The Larva (The Problem Plays)—Much Ado About Nothing, Troilus and Cressida, All's Well that Ends Well, Measure for Measure

❖ 3. The Pupa (The Tragedies)—Hamlet, Othello, Macbeth, King Lear.

❖ 4. The Imago (The Last Plays)—Pericles, Cymbeline, The Winter's Tale, The Tempest)

In Shakespeare, the creative process can be seen as a fourfold helix associated with the parallel processes of metamorphic evolution. For Henry James, the culmination and cessation of Shakespeare's artistic career in The Tempest is a mystery whose power to torment us intellectually seems scarcely to be borne:

What manner of human being was it who could so, at a given moment, announce his intention of capping his divine flame with a twopenny extinguisher, and who then, the announcement made, could serenely succeed in carrying it out? . . .[It] puts into a nutshell the eternal mystery, the most insoluble that ever was, the complete rupture, for our understanding, between the Poet and the Man.[2]

But Shakespeare clearly sensed he had completed his series of fourfold helices. It was not a conscious thing, but so perfectly natural and spontaneous was Shakespeare's art that it had reached a perfect conclusion over the vast canvas of his overall development. Schelling grasped this when he wrote that Shakespeare never portrays either an ideal or a formal world but always the real world. But when he goes on to say that the ideal element manifests itself in the construction of his plays, Schelling has turned the world upside down positing the ideal for the real, since the construction unconsciously conceals an entirely natural and material fourfold significance. As James implies, Shakespeare the ordinary citizen then wanted release from the exigencies of his art:

Here at last the artist is, comparatively speaking, so generalized, so consummate and typical, so frankly amused with himself, that is with his art, with his powers, with his theme, that it is as if he came to meet us more than half-way, and as if, thereby, in meeting him, we were nearer to meeting and touching the man.[3]

This was artistic resignation, or death, the dramatic poet earthed. Henry James's hand on his deathbed was still making the motions of writing with a pen. But Shakespeare had already achieved what James and the novelists continued to strive for, "some copious equivalent of thought for every grain of the grossness of reality . . . the joy of sovereignscience."[4]

George Steiner's pontification is quite erroneous when he pronounces that Shakespeare's manifold and secular humanity is unreceptive of systematic unification, which echoes Schelling's criticism that "he is too diffuse in his universality."[5] The spontaneity and naturalness of the dramatist seems to annoy the critical theorists. Again Steiner writes, irritably: "in most writers, Shakespeare representative among them, the compositional process seems to show no correlations with what we know of the methods of discovery in mathematics. But in some poets (Poe or Valéry, for example), as in musicians, painters or architects, the affinity to mathematical means and ideals is significant. They feel, they construct more geometrico."[6] The basis of medieval psychology has already broken up by the time of Shakespeare, the structure of the four humors set upon the four elements absorbed, but the whole is transformed and extended in relation to personality almost beyond recognition. So, for example, Ben Jonson's plays, despite all their dexterity and wit, plod in comparison, tied to an already surpassed concept of temperament. Shakespeare's plays embody in their full span a scientific apprehension of the creative mind grounded in an unrivalled understanding of human motive and action. What remains is to link it with the universe.

Heidegger infamously remarked of the fullest body of knowledge about the world around us: science does not think. Much of science merely records, however elaborate the theoretical template required or complex the experimentation. Imagination is not so directly bounded by empirical fact. It lingers on the edge of consciousness, only activating when it can hitch a ride on passing fragments of intuition and go on to assemble newfound connections into an entirely new, scientific or poetic, form. Newton famously transmuted the falling apple in his Lincolnshire garden into knowledge of the gravitation forces applying through space and thus to the orbit of the moon, and this is clearly an act of imagination followed up by a formidable analytical intelligence. Then again, centuries later, Einstein in a staggering feat of imaginative re-assemblage firstly conceived a constant velocity for light, and then reassessed the whole of contemporary physics, though today whether the fixity of the speed of light is indeed an absolute is in dispute.

Imagination is connected to a primal human impulse just below the threshold of consciousness, and at its highest philosophical and poetic regions relates to inner reality rather as Minkowski's four-dimensional math relates more exactly to the external world than immediately accessible and commonsensical three-dimensional concepts. This does not prevent some scientists from rejecting the role of art, just as some philosophers and critics downgrade science. So Paul Dirac believed what he called the pretty mathematics lying behind reality rendered the poetic imagination obsolete:

I do not see how a man can work at the frontier of physics and write poetry at the same time. They are in opposition. In science you want to say something nobody knew before, in words which everyone can understand. In poetry you are bound to say something that everybody knows already in words that nobody can understand.[7]

Thoreau had put the counter argument a century earlier:

I should say that the useful results of science had accumulated, but that there had been no accumulation of knowledge, strictly speaking, for posterity; for knowledge is to be acquired only by a corresponding experience.[8]

This mid-nineteenth century statement, though, has been comprehensively rendered obsolete by twentieth century science. Einstein put the matter plainly in relation to modern gravitation theory:

> no ever so inclusive collection of empirical facts can ever lead to the setting up of such complicated equations. A theory can be tested by experience, but there is no way from experience to the setting up of a theory.[9]

However even though the knowledge available through scientific discovery can lead to changes in the behavior of organisms, there remains a question as to whether in the last resort this is any more than a higher form of information.

Even if this is the case, poetry has no cause for complacency. Among contemporary poets only Miroslav Holub has successfully poetically engaged with science, though earlier Paul Valéry had characters debate Einstein in his L'Idée Fixe. A nineteenth century poet has recently been shown not merely to have foreseen Darwinian natural selection, but to have incorporated into his poetry, structures of thermodynamics only in the process of being discovered by Lord Kelvin and others. I mean Tennyson, and his epic elegy, In Memoriam, published in 1850 but gathering together poems written over a period of 17 years. It is well known that the sections of this poem regarding evolution had been circulating for several years prior to Robert Chambers' limited but pioneering Vestiges of the Natural History of Creation of 1844. What has been less well appreciated is that, as an article by Barri J. Gold revealed, In Memoriam is saturated with the language of energy physics, so much so that she concludes that Tennyson can be said to have discovered— poetically—not only the terms but also the principles and processes of the nascent science of energy physics, especially the poetic evocation of the tension between conservation and dissipation that haunts the first and second laws of thermodynamics.[10]

In other words, a truly great poet can intuit advances in science, and indeed scientists of the mid-nineteenth century consulted Tennyson in regard to their latest work. A parallel situation for a poet is almost inconceivable in the Western world today, yet poetry will not reclaim the intellectual and cultural high ground beyond the national and ethnic into which it has stumbled until it recurs.

It is significant that Miroslav Holub has endorsed experimental psychology's conclusion that the present moment lasts three seconds. This is a piece of knowledge that is provocative to the imagination. For if experience and reaction to immediate events appertain to the triple, then we might expect reflection, creativity and religious systems together with the natural systems to which they unconsciously refer to be grounded in the fourfold. Etymology in relation to entomology provokes an interesting conundrum in regard to the issue at stake, and disproves Heidegger's contention that "in order to be who we are, we human beings remain committed to and within the essence of language, and can never step out of it and look at it from somewhere else."[11] Language can propose a higher metamorphosis. It does of course provide a selective advantage to the human species, and as such has a dialectical relationship with genes and environment. But it also offers key words carrying a 'message' that require interpretation. So William Blake's Los is eternally reconstructing language for its inner meanings, empowering symbols to enable the individual to view all four sides simultaneously, imaginatively.

One of the associative components of the word 'imagination' is imago an image, or more to the point here, the inclosed image in the caterpillar that comes forth as the adult of an insect. If the associations of imago are traced back through language, then Latin 'pupa'—the root of pupil—is the chrysalitic stage of insect development, with something of the character of krusos, gold, in the root of the associated process, the 'chrysalis'. Pupa also relates to the pupil of the eye, the twinkling of a thought or theory long before the full work of realized imagination. As Steiner records, the comparison of the pupil of the eye to a small child (pupilla), has been traced in all Indo-European languages, but also in Swahili, Lapp, Chinese, and Samoan.[12] Tunneling further down language takes us to a fundamental word, 'larva', which signifies in the Latin both person and mask. So when Descartes writes "larvatus prodeo"—I proceed masked—the condition of the insect becomes a metonymy for a certain stage of the human condition, as observations may prepare the way for theory through the use of applied imagination. Einstein's remark to Heisenberg sent the latter on the trail of the Uncertainty Principle: It is the theory which decides what we can observe.[13] The caterpillar masks the perfect imago, embryonically contains its enzymes and cells, so likewise the radix of the person is the Latin persona, or actor's mask. By way of these etymological roots, the imagination becomes the true measure of the growth of the conscious human, while at the same time it relates us to the entomological, by way of the succession ovum-larva-pupa-imago. The human species has evolved from the great apes, but the inner labyrinth of language connects us to the insects.

I am proposing that this underground river of meaning ultimately creates a four-dimensional ontology for language and the arts, isomorphic to the four-dimensional ontology of perduring objects favoured by the special theory of relativity. Both go beyond inertia systems. Neither are immediately accessible to referential understanding because of the dynamics of shifting co-ordinates. The totality of the course of a symphony is to be understood as an abstract formulation like Minkowski's concept, a mathematical analogy breaking up the fixity of place time. The next chapter shows how the music of Beethoven, Mozart, Brahms or Bruckner disperses our everyday self, and then dramatizes in musical notation and structure a metamorphic progression which it coaxes out from behind the unconscious and mask of that self. Worldlines or worldworms of personal development become, for the extent of the symphony or quartet, akin to string bundles of psychic events stretched out in the space-time of the unconscious. Through the great works of the imagination, the individual travels up and down the passages of self, undergoing reversion and regression as well as progression. A scientific treatise demands a point-by-point understanding of the stages of the proof or argument, but the works of the imagination immediately penetrate a core of intellect even before what lies behind them can be unraveled. They transform the inner life of an individual by a type of ghostly, abstract experience which is the work of art. Ultimately imagination is an intellectual tool, whereas knowledge is a product of the use of that tool. Knowledge is what the imagination sloughs off in its own capacity as what Nabokov calls the muscle of the soul.

Holub's emphasis on the significance of the three-second moment can now be recognized for the brilliant insight it is. It establishes the limits of the threefold in the world of the intellect, belonging to what Valéry called moi no. 2, or the adaptative self acting under the aegis of experience. The imagination, though, can operate in the quaternal sphere and is a way of identifying the components that can lead to new knowledge. It does this by discovering similarities between things whose likeness had not previously been realized in thought. Indeed the great works of imagination in music and literature can illuminate the now that so exercised and worried Einstein. This is because these works move through time, and yet being abstracted from real time, reveal things that are apparently beyond time. The structures they are built on relate directly to the reader/observer who is the ultimate co-ordinate of now. Whereas Relativity theory places all time on one curve so that past, present and future are contemporaneous, Quantum physics makes time entirely uncertain, to the degree that an atom does not know its next move. Indeed the superposition principle enables a particle to be in two places simultaneously.

Here Shakespeare's poem "The Phoenix and the Turtle" stands as a clear anticipation of the ambiguities of quantum:

> So they loved, as love in twain
> Had the essence but in one;
> Two distincts, division none:
> Number there in love was slain . . .
>
> Property was thus appalled,
> That the self was not the same;
> Single nature's double name
> Neither two nor one was called.

But neither Relativity nor Quantum mechanics, nor even the isolated poem, can explain the human, subjective, sense of the flow of time. Einstein recognized this when he asserted "there is something essential about the 'now' which is outside the realm of science, the experience of which has a special significance for humans." The four-dimensional continuum results in 'now' forfeiting its objectivity in the spatially extended world, and so is left with only a 'frame-relative' existence".[13] Then Gödel goes on to define reality as consisting of an infinity of layers of the 'now' which come into existence successively.[14] Which is why for the most part Shakespeare wrote poetic dramas which could enact this multiple and paradoxical experience.

CHAPTER 9

RE-CONSIDERING ALFRED RUSSEL WALLACE'S DISTINCTIVE CONTRIBUTION TO EVOLUTIONARY THEORY

A.R. Wallace had a recurrent dream during his early childhood in the Usk Valley "as of some creature with huge wings." He ultimately traced its origin to a funeral escutcheon he had seen, and which had played on his imagination like "an unmeaning jumble of strange dragon-like forms."[1] Prophetic that his most memorable experiences as a naturalist should involve beasts with huge wings! For central to Wallace's life would be the insects, and specifically the tropical Lepidoptera, butterflies and moths. The structure and coloration of Lepidoptera play a major role in his disputes with Darwin. The thick scale exterior from which the species get its name (lepis = scale) evolved from ancestral hair cover. Indeed moths and the more primitive butterflies (Hesperidae or Skippers) retain dense hair cover at the thorax. The selective advantage of scales for the diurnal moths and butterflies arose from their allowing a cooling cover in the sun, and protection from water loss. Moreover, the bright coloration of scales in certain groups afforded both protection and a signaling ability to find for reproduction partners.

On 23 February 1867, Darwin wrote to his younger colleague:

> On Monday evening I called on [H.W.] Bates, and put a difficulty before him which he could not answer, and, as on some former occasion, his first suggestion was, 'You had better ask Wallace.' My difficulty is, Why are caterpillars sometimes so beautifully and artistically coloured? Seeing that many are colored to escape dangers, I can hardly attribute their bright colour in other cases to mere physical conditions.[2]

Wallace was at the time preparing "my rather elaborate paper" on 'Mimicry and Protective Coloring' for the Westminster Review.[3] Mimicry shields some organisms that are palatable to predators, by their evolving a similarity to those unpalatable. Wallace explained that a caterpillar's bright colors, by making it so visible, encouraged its enemies to distinguish it at a glance from others attractive to taste. (It has also recently been argued that there are palatable lepidoptera that imitate the movement behaviour of unpalatable prey.)[4] The principle of ostentatious marking holds especially with butterflies of the Papilionidae and Heliconiidae families, which have a strong smell and unpleasant taste protecting them from insectivorous birds and other creatures. When Darwin objected on the grounds of differences in coloration between male and female, Wallace replied drawing on his observations of the lesser mobility and vulnerability of the female during ovipositing (that is, egg-laying):

> Your objection that the same protection would to a certain extent be useful to the male, seems to me utterly unsound, and directly opposed to your doctrine so convincingly urged in 'Origin' that Natural Selection never can improve an animal beyond its needs.

> A male, being by structure and habits less exposed to danger and less requiring protection than the female, cannot have more protection given to it by Natural Selection, but a female must have some extra protection to balance the greater danger, and she rapidly acquires it one way or another.[5]

Already in a paper of 1865, Wallace had begun to dispute the territory which Darwin put outside the laws of natural selection. He shows there how the more subtle camouflage of female butterflies is occasioned by "their slower flight laden with eggs."[6]

Mimicry was becoming a key issue with evolutionists even as Darwin published his The Origin of Species in 1859. Wallace declared that Darwinian principle "leads us to seek an adaptive . . . purpose . . . in minutiae which we should otherwise be almost sure to pass over as insignificant or unimportant."[7] Wallace pointed out that the two sexes of Papilio paradoxa exactly imitate Euploea midamus chloe, and appear in the same districts. So similar are they that he "could hardly ever distinguish them on the wing. The male of the Euploea "has the fore wings of a brilliant metallic blue, with faint bluish-white spots, while the hind wings are uniform brownish black. The female differs considerably, the hind wings being covered with narrow white lines radiating from the body, and having a marginal row of white spots."[8] We should hardly be surprised that mimicry plays a role in organic nature. After all it is common in the world of physical matter. Every newly created particle is accompanied by a sort of 'negative image' partner, a so-called antiparticle somewhere in the universe however distant. So an electron (which carries a negative electric charge) is always created along with an antielectron or positron, which has the same mass as the electron but an opposite, positive, charge. Every proton is shadowed by an antiproton. This doubling process is a feature of nature, and is imaginatively built upon by many writers from Hogg to Nabokov.

The arguments between Darwin and Wallace became sharper with time, as Darwin dug his heels in, insisting that the brighter wings of males, whether bird or butterfly, was primarily as an attractor to females. Since butterflies respond to color when courting, sexual selection resists color change in males. Hence the male swallowtail retains an ancestral color pattern. When in The Descent of Man, Darwin writes that "the colors of caterpillars are mostly protective, being due to natural selection alone, while those of butterflies are mostly attractive, being largely due to sexual selection," Wallace answers in 1879 that this is a slurring over of "what is really a stupendous difficulty in the way of the theory. So far from the colors of caterpillars being 'mostly protective' every entomologist knows that a large number of caterpillars in every part of the world are conspicuously colored, and what is more to the point that their colors are as brilliant and varied as those of butterflies themselves, if we take into account the nature of their integument, the small amount of surface, and the uniform cylindrical form of their bodies."[9] What Darwin had called 'minor' causes of modification—immediate conditions, use and misuse, variability and habit together with more sophisticated secondary traits of beauty—had for Wallace to come within the purview of the necessities of natural selection. Martin Fichman puts the matter succinctly:

> To Wallace, evolutionary explanations of behavioral or physical traits predicated on an aesthetic sense in lower animals were unacceptable. The discontinuity between human higher faculties and the mental processes of the rest of the animal kingdom had become axiomatic for him. Wallace declared that imputing

aesthetic tastes to birds (and insects) was an anthropomorphism as unwarranted as that made by 'writers who held that the bee was a good mathematician, and that the honeycomb was constructed throughout to satisfy its refined mathematical' sense.[10]

Wallace was already becoming what he later christened himself: "more Darwinian than Darwin," which is quite strange since he had such very serious cultural and spiritual differences with the author of Origin of Species.

There is no mimicry among butterflies in Britain, though it does occasionally occur among moths, as with the Broad-bordered Bee-hawk and Hornet Clearwing. (The lines of Robert Browning, which especially appealed to Nabokov, play on this lepidopteral feature—"lichens mock/ The marks on a moth, and small ferns fit|Their teeth to the polished block.") Abroad though there are the Papilionidae considered, as Wallace put it, "by all the older writers to be the princes of the whole lepidopterous order," and "with opalescent hues, unsurpassed by the rarest gems."[11] Above all for the student of evolution, "they exhibit, in a remarkable degree, almost every kind of variation, as well as some of the most beautiful examples of polymorphism and of mimicry," along with a wide variety of localization.[12] As Wallace remarks in his article on "The Malayan Papilionidae or Swallow-Tailed Butterflies, as illustrative of the Theory of Natural Selection," these insects offer "immense development and peculiar structure of the wings which not only vary in form more than those of other insects, but offer on both surfaces an endless variety of pattern, coloring, and texture."[13] Moreover the prevalence of mimicry was extensive among the swallowtails: "In all parts of the world there are certain insects which, from a disagreeable smell or taste, are rarely attacked or devoured by enemies. Such groups are said to be 'protected,' and they always have distinctive and conspicuous colors."[14] Their mimics can be of either sex, "but most frequently it is the female only that is thus modified, especially when she lays her eggs on low-growing plants; while the male, whose flight is stronger and can take care of himself, does not possess it, and is often so different from his mate as to have been considered a distinct species."[15]

It was Wallace, rather than H.W. Bates (or Fritz Müller), who defined the pre-eminence of colour's role as the key to warning predators, rather than smell or any other characteristic. So in his 1866 paper on "Natural Selection," he writes:

1. In all cases of mimicry, the resemblance of the one species to another in a different group is entirely superficial, and is always strictly confined to those characters which cause the one to look like the other. [This is confirmed by work on mithochondic evidence on mimic butterflies and geography] The structure, the habits, the form of inconspicuous parts, the nature of the food, or the character of the larva and pupa, are not, as far as we know, ever modified in a similar manner . . . .

2. There are no grounds for believing that minute details of colouration and marking are due to climatal conditions at all, still less that they can be produced so identically alike in species of groups widely differing in organization . . . ."[16]

A section of Malay Archipelago gives another angle on Wallace's observations on mimicry. Once again, they are very precise, and emphasize his intimation that the Lepidoptera have a special role in the theory of evolution. The handsome Papilio memnon is "a splendid butterfly of a deep black color, dotted over with lines and groups of scales of a clear ashy blue" with wings "five inches in expanse, and the hind wings are rounded, with scalloped edges. This applies to the males; but the females are very different and vary so much that they were once supposed to form several distinct species. They may be divided into two groups—those which resemble the male in shape, and those which differ entirely from him in the outline of the wings." One of the females led Wallace "to discover that this extraordinary female closely resembles (when flying) another butterfly of the same genus but of a different group (Papilio coön)." This resemblance occurs because "the butterflies imitated belong to a section of the genus Papilio which from some cause or other are not attacked by birds, and by so closely resembling these in form and colour the female of Memnon and its ally, also escape persecution."[17]

Richard Dawkins has asserted there is a supergene that effects the approximation of the imitator to the model butterfly.[18] This does not now seem to be the case. Instead in the course of time the approximation becomes perfect resemblance. This occurs as a gradual process of natural selection; an imitation that becomes closest in appearance to the unpalatable organism has a greater chance of escaping the predators, and this imitation over generations becomes the norm by way of genetic drift. The bio-chemical process is created by a morphogenic diffusion that stimulates a gene or color-specific enzyme to generate a stable spatial pattern. Or in another type of mimicry where a butterfly or moth may resemble a leaf, "Even the peculiar colors of many animals, especially insects, so closely resembling the soil or the leaves or the trunks on which they habitually reside, are explained on the same principle [as the giraffe's neck]; for though in the course of ages varieties of many tints may have occurred, yet those races having colors best adapted to concealment from their enemies would inevitably survive the longest."[19] In "On the Law Which Has Regulated the Introduction of New Species" of 1855, Wallace concluded that "every species has come into existence coincident both in space and time with a pre-existing closely allied species." So new species arise through variations which enhance survival: "The superior variety would then alone remain, and on a return to favorable circumstances would rapidly increase in numbers and occupy the place of the extinct species and variety. The variety would now have replaced the species, of which it would be a more perfectly developed and more highly organized form."[20]

It is infamous now that, despite its title, The Origin of Species does not centrally address the problem of speciation. As Wilma George put the matter: "It was concerned with establishing natural selection as the directive force in the steady evolution of one species into another over geological time. Little was said about the circumstances in which a population could be divided into two or more groups each of which would become a species contemporaneously."[21] Wallace spelt out the conditions under which speciation to take place:

> That whatever the amount of variability of a species, no general modification of it will occur so long as the environment remains unchanged; and (2) that when a permanent change (not a mere temporary fluctuation) of the environment occurs—whether of climate, of extension or elevation of land, of diminished food-supply, or of new competitors, or of new enemies—then, and then only, will various specific forms become modified, so as to adapt them more completely to the new conditions of existence . . . .

> In order to be developed through natural selection a particular variation must not only be useful, but must, at least occasionally,

be of such importance as to lead to the saving of life, or to use Professor Lloyd-Morgan's suggestive term, be of 'survival-value.'[22]

Speciation remains a hotly contested issue. It was raised by E.B. Poulton in "What is a species?" His species concept elucidated in that 1904 address is, as James Mallet explains, that they are "syngamic (i.e. formed reproductive communities), the individual members of which were united by synepigony (common descent). Poulton's species concept was informed by his knowledge of polymorphic mimicry in Papilio butterflies: distinct non-mimetic male and mimetic female forms were members of the same species because they formed syngamic communities."[23] Significantly Wallace had just given Poulton a book on mimicry in December 1903, which included the first mimicry papers, by himself, H.W. Bates and Roland Trimen.

So the study of mimicry is at the heart of evolutionary theory, and it is again interesting that the world of insects in which Wallace had immersed himself from an early age provides the key to these issues. A core of the argument centers on structure, and by implication even upon the nature of language. In overthrowing Paley's argument for the existence of God—which might agnostically be read as Perfection—from design, Darwin simultaneously threw out most of the aesthetic sense, the baby with the bathwater, except in the instance of sexual selection of mates. Significantly Wallace's much-quoted rhapsodic response to Ornithoptera croesus had already suggested an almost aesthetic function for lepidoptera. When Darwin writes in his Autobiography, "we can no longer argue that, for instance, the beautiful hinge of a bivalve shell must have been made by an intelligent being," the premise is rightly the fixity of the laws of nature.[24] But in this epoch of a far greater knowledge of internal mechanisms, it is now possible to unify structural design and nature's systematizations. This is not to return to the Lamarckian belief in the existence within organisms of a built-in drive toward perfection as part of the heritability of acquired characteristics. Darwin's discoveries on speciation rejected the latter and, as the neo-Darwinians insist, speciation is usually the consequence of the divergence of populations separated by a geographical divide. (It was Ernst Mayr who was to formulate the species concept based on reproductive isolation). This is also the difference between Lamarck's vertical and Darwin's horizontal time; the latter is concerned with the origin of the diversification of species, and dismisses Lamarck's model of groups of organisms progressing independently by spontaneous generation towards perfection. So selection is the outcome of variation, and mistake rather than deliberate design is the pivot. During DNA copying sequences in reproduction, errors enter. The resultant copy is imperfect, and the proteins coded differ. Evolution occurs through such discrepancies. Far from striving for perfection, the root of the variations from which natural selection produces evolutionary change is such error in genetic transmission. So it is the concept of process that reunifies design and fixed laws.

In regard to further differences between Darwin and Wallace, A.J. Nicholson has defined a variation of emphasis between Darwin's 'competitive selection' and Wallace's 'environmental selection.' Under competitive selection, "the individuals that do not survive are not really unfit at all. They are simply less fit than the survivors and are eliminated only because of the presence and preservation of the more fit individuals."[25] For Wallace though it was the external environment that set an absolute challenge to be met in order to survive so that, as he expressed it, "if there were "some alteration in the physical conditions" such as an infestation of locusts extinguishing a parent species, "the variety would now have replaced the species, of which it would be a more perfectly developed and more highly organized form."[26] In time the variety would become the new species. At the same time, Darwin was aware of both factors even though he continually turned to the arguably more societal argument. Significantly, Wallace had no interest in artificial selection and domestication, and always took his examples from natural populations. This, however, meant he failed to explain the divergence of species. "Every species comes into existence coincident in time and space with a preexisting closely allied species," he wrote in Annals and Magazine of Natural History through which he first came to Darwin's notice. So he described the fact but not the mechanism of divergence. Or as Peter Bowler has it, "Wallace simply assumed that species split into varieties— he did not seek to explain how this all-important first step occurs."[27] The process could only be explained by those societal causes that Wallace shunned, for it is diversity which reduces competition between organisms until such a point as divergence of species occurs. Perhaps the best definition of speciation remains that of Karl Jordan produced in 1896 while employed by Walter Rothschild, and favored by Mallet:

"A species is a group of individuals which is differentiated from all other contemporary groups by one or more characters, and of which the descendants which are fully qualified for propagation form again under all conditions of life one or more groups of individuals differentiated from the descendants of all other groups by one or more characters."[28]

But on a basic question such as the reason for the upright posture of humans, Wallace was able to provide the environmental answer that was lacking in Darwin. His adaptationism undoubtedly led him to over-estimate utility as an all-pervading factor, but at the same time it did mean that he could not accept the relativism with occasional lapses into fundamentalism in regard to artistic and spiritual matters, allowed by the pluralism of Darwin.

It was in 1869 when Wallace reviewed the tenth edition of Lyell's *Principles of Geology* that for the first time he stated publicly and categorically his belief that natural selection could not account for the growth of the mind of humans. Although it "may teach us how, by chemical, electrical, or higher laws, the organized body can be built up, yet it "cannot be conceived as endowing the newly arranged atoms with consciousness."[29] Even as early as 1864—the date of The Origin of Species was 1859—Wallace had insisted that the whole of natural selection had changed with the evolution of the human brain. But Darwin wrote in one of his Transmutation Notebooks enquiring "why is thought being a secretion of the brain, more wonderful than gravity a property of matter? It is our arrogance, our admiration of ourselves."[30] What would he have made of Gerard Manley Hopkins's description of Purcell's music as having "uttered in notes the very make and species of man as created both in him and in all men generally"?[31] Darwin's Autobiography attests to an outlook that has led to many of the cul-de-sacs in contemporary thinking: "My mind seems to have become a kind of machine for grinding general laws out of large collections of facts, but why this should have caused the atrophy of that part of the brain alone, on which the higher tastes depend, I cannot conceive."[32] This Gradgrindery was the root of Darwin's later inability to react to music and poetry, other than his beloved Milton, though novels being a product of "not of a very high order" of imagination were read for relaxation.[33] The primary discoverer of the laws of evolution himself becomes crucial evidence in favour of Wallace's reservations as to the universality of their application. In the course of his long life, Wallace came to feel the need for a more comprehensive view of the human situation, of humanity's relation with nature, than evolutionary theory allowed. The spiritualism he came to later in life was part of that. Certainly Wallace evinces a deeper feeling for nature than Darwin, not least in his response to the variety of lepidoptera. Again, the Papilionidae which include most of the great Birdwings, offered much for the theoretician as well as the naturalist and even aesthete: "the family presents us with examples of difference of size, form, and color, characteristic of certain localities, which are among the most singular and mysterious phenomena known to naturalists."[34] And when he saw the Ornithoptera poseidon, he experienced "the joy which every discovery of a new form of life gives to the lover of nature"—"It is one thing to see such beauty in a cabinet, and quite another to feel it between one's fingers, and to gaze upon its fresh and living beauty, a bright-green gem shining out amid the silent gloom of a dark and tangled forest."[35]

As we have seen, Wallace maintained that the human brain specializes in "the evasion of specialization."[36] He viewed this evasiveness as characteristic of the higher evolution of humans. The brain, after all, is an intricate interrelation between over 100 billion cells and even at the merely physiological level, the molecular events accompanying thought and its attendant, memory, are hardly conducive to the banal fixities of one-sidedness. So Sir John Eccles asks whether there is "some process that we could call genetic dynamism whereby the hominid brain inevitably develops further and further beyond natural selection?"[37] The configurations of both natural and artistic laws in their spontaneous symmetry strike us as beautiful, and present themselves today with the urgency Wallace expressed over a century ago when writing of the Bird-winged tropical butterflies and the Great Bird-of-Paradise. Humanity, he warned, threatened so to disturb the balance of nature as to cause the extinction of those creatures whose "wonderful structure and beauty" it alone is in a position to appreciate. Animals and insects on the other hand were determined by particular demarcation, whether in a leopard's sense of smell or a butterfly's dependence upon a single food plant for its larval phase. Wallace argued time and again, despite his rigorous expositions of Darwin's theory, that to fail to understand this is to open the way for cultural regression. The last chapter of his *Darwinism* (1889) almost forgets the first 14 chapters of uncontroversial exposition as he strives to illuminate the distinctively human quality of the brain: "Because man's physical structure has been developed from an animal form by natural selection, it does not necessarily follow that his mental nature, even though developed pari passu with it, has been developed by the same causes only."[38] He goes on to argue that mathematics, music and art "clearly point to the existence in man of something which has not derived from his animal progenitors—something which we may best refer to as being of a spiritual essence or nature, capable of progressive development under favorable conditions."[39] And he draws attention to "the workings within us of a higher nature which has not been developed by means of the struggle for material existence."[40]

Wallace's advocacy of the cultivation of this "higher nature" had led to a serious cooling of relations with Darwin later in their careers. But research a century or so later provides a clue to Wallace's accuracy. For, as Richard Dawkins has argued, natural selection operates at the level of competing genes, not competing organisms. It is genes that are engaged in a struggle for existence, and they therefore do everything they can to increase reproduction in the next generation. This explains altruism; the sacrifice of one individual ensures the survival of other family or tribal members, or as Dawkins seems unable to accept, the species as a whole in the instance of the International Brigades volunteers in the 1930s, or Byron's volunteering to assist in the Greek's War of Independence in the 1820s. Stephen Jay Gould disagreed with Dawkins on this point, and it seems contemptuous of human thought-processes to reduce the species to "gigantic lumbering robots" which harbor gatherings of the great replicator, the gene that is "the basic unit of selfishness." According to Dawkins "we were built as gene machines, created to pass on our genes."[41]

The evolutionist, Loren Eisley, has sharply crystallized the situation of Wallace's reservations on the relation of art to evolution: Darwin "did not, however, supply a valid answer to Wallace's queries. Outside of murmuring about the inherited effects of habit—a contention without scientific validity today—Darwin clung to his original position. Slowly Wallace's challenge was forgotten and a great complacency settled down upon the scientific world."[42] The constant irritation for Darwin was that his younger colleague often anticipated him, and even threatened to surpass in theoretical boldness his own perspectives. It had become clear to Wallace that the super-structural elements in culture play some independent role, even if in his questing spiritualism Wallace never became entirely clear what that may be. Since then many scientists have reduced Darwin's work to a form of nihilism. Indeed when Gould writes triumphantly that "Darwin cut through 2,000 years of philosophy and religion in the most remarkable epigram of the M notebook: 'Plato says in Phaedo that 'our imaginary ideas' arise from the preexistence of the soul, are not derivable from experience—read monkeys for preexistence,'" Darwinism has been transformed to absolutist reductionism.[43]

CHAPTER 10

MORPHOLOGY AND MORPHEMES

The molecular biologist, Sean B. Carroll has summed up the growing knowledge of human's insect debt as a result of research over the past quarter century:

> "Morphologically disparate and long-diverged animal taxa share similar toolkits of body-building and body-patterning genes. The best known discoveries of evo-devo [i.e. 'evolutionary developmental biology'] are those concerning the presence of homeobox-containing Hox genes and Hox gene clusters in flies, vertebrates, and most animal phyla. The similarities in animal genetic toolkits extend to a large number and wide variety of transcription factors and components of signaling pathways. The phylogenetic [i.e. branching out from the common ancestor] distribution of toolkit genes suggests a fairly complete modern toolkit was in place in the last common ancestor of bilaterians, prior to the Cambrian period. [544-490 million years ago]"[1]

The great differences in appearance and behavior between insects and humans is fundamentally deceptive. Actually the divergences in body forms are regulated by parallel groups of genes. Even more paradoxically, similar signaling molecules are found in yeast and plants predating mammal life. It is the expansion and complexity of the vertebrate nervous system that has created the modern human, and allowed the human species its dominant position in nature. But even so, "the morphological diversity of vertebrates from humans to hummingbirds, or from whales to snakes, evolved around a common set of developmental genes."[2] All share the same set of determining 39 Hox genes.

Language also tells the human story in a complex, but clear linguistically-historical manner. It parallels the genetic strands, but has arisen through the interplay of language development and the natural world. This is a tale told in Latin, but the philosophical and linguistic roots of the story lie in the Greek—philosophically in Plato's Phaedrus and linguistically in the word ψυχή.[3] Psyche is a key word in Greek since it evolves from Homer to signify not only 'breath of life,' but also mind and soul, and is ultimately imaged as a butterfly. The of the Greek word psyche, ψυχή, is intriguing. The letters ψ (ps) and χ (kh) have an interesting history. They are two of the three so-called supplementals added by the Ancient Greeks to the Canaanite Semitic script adapted from a syllabic language to create the first alphabet which we in the West have, of course, inherited. φ, χ and ψ do not appear in Phoenician, and were added by the Greeks.[4] The supplementals belonged to the earliest alphabet, but at that stage ψ represented the sound 'kh'. The first set of supplementals arrived among the Euboians. Later, a reformer, an Ionian, clarified χ as kh and ψ as ps, discarding φσ as 'ps' and χσ as 'ks.' While Athens retained these older forms, Corinth took up the new symbolic significances. Later the modified letters became part of the Koine script. So these were aspirates which came to form the core of the word ψυχή, itself signifying, inter alia, 'breath.'[5] Eric Havelock sees invention by the Greeks of the alphabet as dissolving "the syllable into its acoustic components—we might almost say its biological components in so far as these are actually effects produced by movements of different parts of the human body."[6] Bickel argued that the Homeric ψυχή did indeed represent the very process of respiration.

Barry Powell argues, following H.T. Wade-Gery, that "the signs ψ and χ introduce confusion, not clarity,"[7] and yet here he forgets his main thesis that the literature of Greece produces the language through the adapter who wrote down the hexameters of Homer. (Havelock's term 'inscriber' for the adapter is a more appropriate term.)[8] Powell insists it is not at all likely the three supplementals were invented as a result of evolutionary need.[9] However Andrey Bely offers a different perspective on this. In his essay "The Magic of Words," Bely points to the role of literature in creative cognition:

When I assert that creation precedes cognition, I am asserting the primacy of creation, not only because creation is epistemologically superior, but also because it is prior in actual genetic sequence.[10]

So in the two signs under discussion, subjective intuitions of the Greeks inspired their invention—intuitions that finally reached the arena of literature in the Latin tale of Cupid and Psyche by Apuleius, which had been intimated in the Rig Veda.[11] When Powell writes that ψ and χ will have belonged to the original system because "we would expect new signs added to a preexisting signary to clarify ambiguity," and that "the signs ψ and χ introduce confusion, not clarity," he is making assumptions about the rationality of alphabetic introductions that misunderstand the "genetic sequence" to repeat Bely's phrase.[12] Letters of the alphabet were defined by the Greeks as 'voices', that is to say that the sound was primary. Indeed Powell makes the pertinent point that Archaic Greek "in its obsession with phonetic accuracy . . . was a great anomaly in the history of writing."[13] As W.B. Stanford put it: "Written words were more like memory-aids to remind readers of certain sounds . . . The Greek grammarians emphasized 'orthoëpy,' the correct rendition of texts "according to intonation, timbre-quality, and quantity, as the prerequisite of literary appreciation."[14] So these supplementary signs were necessary to the Greeks as elements to be assembled for the concise expression of their concept of the soul, which became a form of mimesis through the breath sound of ψυχή. This is in line with the remarkable scientific accuracy of their alphabet:

The inventors of the atomic theory of matter were the first possessors of a system of writing where graphemes represent the 'atoms of spoken language,' an analogy explicit in the Greeks' use of the word stoicheion, 'something in a row,' both for an alphabetic sign and for an atomic element. Greeks imagined that the structure of their writing paralleled the structure of the phenomenal world, according to the unobvious theory that matter consists of a limited number of discrete particles invisible but real, which act in combination to produce visible effects.[15]

The Indo-Europeans generally depended on syllable counting; the Greeks on length, short and long. Boisacq distinguishes a provisional Indo-European etymology in *bhes- (souffler) leading onto ψυχή and Bickel interprets the Greek word as "breath-soul".[16] On the other hand Claus insists throughout his book Toward the Soul: an inquiry into the meaning of ψυχή before Plato that it not so much signifies 'breath' as 'life-force' together with an intimation of coldness. However there is an unusual confirmation of the origins of the nuances of the word. Recent medical research has been carried out into the relation between reading Homer aloud and cardiovascular health:

The effects of different breathing frequencies and patterns found in poetry readings on cardiovascular regulation have been investigated extensively in recent years. Poetry recitation has been known to cause a frequency adjustment of breathing oscillations with endogenous blood pressure fluctuations (Mayer waves) and even cerebral blood flow oscillations during the saying of the Catholic Rosary and the 'OM' mantra.

The effect is attributed to the breathing frequency of approximately six breaths per minute induced by the metric of both religious verses . . . so that heart rate and respiration may intermittently synchronize."[17]

The product of these originally oral poems of Homer consists both of an impact of poetic thought and at the same time a guarantor of actual health. The result is utilitarian and aesthetic, as well as an affirmation of the ancestral traditions of the community of listeners and speakers. The long and short syllables of the hexameters of the Iliad and Odyssey are particularly beneficial. During recitation of hexameter verse the low frequency oscillations of the breathing pattern were synchronized to a large extent with the heart rate oscillations. The six meters or rhythmic units per line slows the normal breathing rate from about 15 breaths a minute to just 6. This in turn synchronizes with the regular fluctuations of blood pressure which normally go in ten-second cycles. Powell asserts that the dactylic hexameter "is controlled by a structure 'deep' in the user's psyche."[18] Perhaps in the light of the recent scientific research, 'diaphragm' could be substituted for 'psyche' in this instance. Chorus and audience would recite more than 10,000 lines without pausing at performances in Ancient Greece.

Bruno Snell has pointed out that Homer's uses of psyche are not individuated in the modern way, that the psyche in no way belongs to the person. Achilles' reactions "are not explicitly presented in their volitional or intellectual form as character, i.e. as individual intellect and individual soul. Mental and spiritual acts are due to the impact of external factors, and humanity is the open target of external forces which impinge on him, and penetrate his very core."[19] This receptivity to the outside impressions again gives an added objectivity to the ψυχή, which will later in history take on an independent existence as an insect whether a butterfly as in Estonia and Ireland, or bee, wasp or dragonfly in Japan. The soul is in this sense purely objective; it is the breath of life and is quite distinct in life from the person. This objectivity arises from the very apparent weakness in psychological portrayal of Homeric heroes, their failure to distinguish between what was inside and outside themselves as moderns do. At death the psyche departs the individual for Hades, or is loosed. It is this sheer objectivity, the non-personal soul, that makes the Homeric epic so riveting and noble, arousing a sense of truth to life.

The incorporation of the psyche into personal psychology does not take place until the end of the fifth century when a person is conceived as relating to his or her psyche in the course of a life of independently selected activity. In Homer the psyche is a phenomenon in a state of transition. It is likely there was a word for the free soul which in Homer is gradually replaced by the soul as 'breath of life' while at the same time beginning to lose its purely physical limitation. Despite the remoteness of concepts—because of the remoteness—there is an awful power to the Iliad and Odyssey. As Snell puts it:

The psyche [in Homer] leaves through the mouth, it is breathed forth; or again it leaves through a wound, and then flies off to Hades. There it leads a ghostlike existence as the specter (eidolon) of the deceased. The word psyche is akin to ψυχειν, 'to breathe', and denotes the breath of life which of course departs through the mouth; the escape from a wound evidently represents a secondary development.[20]

When Andromache swoons, Homer says that she "breathed forth" (ἐκάλυψευ) her ψυχήν (Iliad, 22, 467), a word "most likely connected with smoke" as Bremmer remarks.[21] All other Greek words connoting aspects of the individual soul are also connected to 'breath'—thymos, noos and menos. The paradox is that it is only at the instant of death that the Homeric psyche becomes evident. During the life of the hero it is implicit, for his identity is determined by his organs. Yet at death psyche has the muscular strength of an organ, an entirely objective entity in no way dependent upon subjective states. This projects the human into a special relation with nature, which is defined as Fate, while the word ψυχή also later reaches across to be imaged as a butterfly. By the time of Hesiod and the lyric poets, it has drawn closer to the psyche element in our own concept of psychological but the concept of the grandeur of the human has begun to shrink.

R.B. Onians though delves into the pre-Homeric origins of ψυχή by way of Plato's Timaeus. This will suggest the physiological roots of the sounds of the word, akin to the snorting that John Burnet rather daringly proposed in his pioneering address to the British Academy some 90 years ago.[22] Onians argues that ψυχή is located "in the marrow, the divine part in the marrow of the head called ἐγκέφαλος . . . The ψυχή is itself 'seed' (σπέρμα), or rather is in the 'seed', and this 'seed' is enclosed in the skull and spine . . . It breathes through the genital organ. This appears to be original popular belief."[23] For Aristotle the seed was itself breath or had breath (πνεῦμα), and procreation itself was such a breathing or blowing.[24]

And so the early psyche is both more and less than the 'breath-soul'. It is not lung-related, but in a sense is more profoundly related to sexual generation. It suggests an erotic root for our word, and this is confirmed by the Dionysiac image on a black amphora from the sixth century BCE which shows a butterfly beneath four drops of semen from a dancing satyr.[25] Then again Hermes was represented as a squared pillar head with genital organs, while an Etruscan scarab shows him with a butterfly on his shoulder.[26] Here the psyche is fertilized with generative powers, so representing the new epoch of wealth-creation centered on craftsmen and merchants, as Norman Brown proposed in his *Hermes the Thief.* And so the circle of ψυχή meanings close, offering life-force together with breath-soul.

ENDNOTES

Forward
[1] Wallace (5), 2:474.
[2] Eisley (1), 306.

Chapter 1
[1] Courcel, 223
[2] Brown, 96-97.
[3] Novalis (2), 53.
[4] Novalis (1), 158.
[5] Novalis (2), 93.
[6] Hiebel, 44.
[7] O'Brien, 112.
[8] Müller, I, 406.
[9] Müller, I, 432.
[10] Müller, I, 440-41.
[11] Müller, I, 357, 377.
[12] MacDiarmid, 55.
[13] Darwin, Chapter 3.
[14] This paragraph owes a great deal to Müller's researches.
[15] Steiner, 125.
[16] Deacon, 115.
[17] Deacon, 265
[18] Ríos (2), 182.
[19] Paz, 55, 40-41.
[20] Chomsky (2), 44.
[21] Wittgenstein, 395-97.
[22] Immisch, 193. These comments by Immisch are followed by this crucial addition:
Als Gesichert wird vorausgesetzt, daB auch Seelenschmetterling eine Abart sogenannten Seelenvogts ist, nichtals das einzige Insekt. Es steht also an Anfang etwas, was weit entfernt ist von dem ammutigen Elfentum der

spätren Faltermädchen und was sonst von zierlicher Symbolik in Frage kommt. Seelenvögel sind ein unheimliches Gelichter.

[23] Schlam, 7.
[24] Fitch, 400.
[25] Havelock (1), 69.
[26] Snell, 20.

Chapter 2

[1] Defries, 241.
[2] Nabokov, 30-32.
[3] Dawkins, 358-59.
[4] Solso, 39.
[5] Kurzweil, 194.
[6] Shubin, 192, 26-27.

Chapter 3

[1] Descartes (3), 213.
[2] Hobbes, 207.
[3] Brown, 90-91.
[4] Rohde, 6.
[5] Budick, 6.
[6] Menn, 5.
[7] Wesson, 22.
[8] Bremmer, 22.
[9] Kahn, 106, 107.
[10] O'Connell, 88.
[11] Kant, 181.
[12] Kimura, 55.
[13] Prud'homme, 468, 769.
[14] Quiring, 265, 785.
[15] Truman and Riddiford, passim
[16] Wigglesworth, 43.
[17] Gilbert, 4.
[18] Rasnitsyn, 458.
[19] Mayr (1), 11.
[20] Mayr (2), 1.

Chapter 4

[1] Jaquette, 2.
[2] Tanner, 46.
[3] Barrow (2), 67.
[4] Schopenhauer (1), 3:325.
[5] Magee (2), 187–88.
[6] Spooner (1), 146–47.
[7] Schopenhauer (3), 1:38.
[8] Pinker, 534.
[9] Pinker, 529.
[10] Altenmüller, 21.
[11] Altenmüller, 26–28.
[12] Schopenhauer (3), 1:81.
[13] Spengler, 1:282–83.
[14] Rosen, 58.
[15] Simpson, 57.
[16] Mellers, 3:92.
[17] Mellers, 3:112.
[18] quoted in Ballantine, 32.
[19] Schopenhauer (1), 3:11–12.
[20] Wenzl, 586.
[21] Schopenhauer (1), 629–30.
[22] Magee (2), 183–84.

Chapter 5

[1] Girard, 53.
[2] Tannery, 115.
[3] Frohock, 10.
[4] Malraux (11).
[5] Malraux (9), 91.
[6] Malraux (8), 60.
[7] Malraux (8), 179.
[8] Malraux (2), 483–84.
[9] Malraux (2), 556.
[10] Carduner, 43.
[11] Malraux (10), 289–90.
[12] Malraux (9), 113.
[13] Malraux (13), 5.
[14] Clare Malraux, 56.

[15]   Bevan (2), 19-29.
[16]   Malraux (9), 91.
[17]   Horvath, 1.
[18]   Bevan (1).
[19]   Malraux (1), 51.
[20]   Courcel, 223.
[21]   Malraux (5), 104.
[22]   Malraux (4), 310.

Chapter 6
[1]   Lindorff, 17.
[2]   Miller, 192.
[3]   Miller, 196.
[4]   Lindorff, 172.
[5]   Lindorff, 140.
[6]   Lindorff, 198-99.
[7]   Meier, 203-04.
[8]   Lindorff, 59.
[9]   Miller, 212.
[10]   Meier, 187.
[11]   Miller, 208.
[12]   Pauli, 420.
[13]   Miller, 212.
[14]   Meier, 203-04.
[15]   Lindorff, 51-57.
[16]   Meier, 146.
[17]   Pauli, 24.

Chapter 7
[1]   Ryanto, 14.
[2]   Husserl (1), III: 254.
[3]   Ryanto, 16.
[4]   Keller, 219.
[5]   Husserl (3), 6.
[6]   Faye, 95.
[7]   Husserl (3), 12.
[8]   Spiegelberg, 92-93.
[9]   Keller, 1-2.
[10]   Keller, 13.

11   Husserl (2), 44
12   Husserl (2), 66.
13   Wrathall, 252.
14   Heidegger, 8.

Chapter 8
1   Keats, 2:67.
2   James, 438.
3   Schelling, 271. with modifications in the translation.
4   James, 432, 431.
5   Simpson, 137.
6   Steiner (3), 73-74.
7   Farmelo passim.
8   Thoreau, 364-65.
9   Schlipp, 1: 89.
10   Gold, 450-51.
11   Heidegger,134.
12   Steiner (1), 102.
13   Einstein, 149.
14   qd. Yourgrau, 164.

Chapter 9
1   Wallace (5), 1:27.
2   Wallace (5), 2:3.
3   Wallace (5), 2:3.
4   Ruxton, 2135.
5   Marchant, 224-25. Wallace's emphasis.
6   Qd. Vorzimmer, 222. In Wallace (18).
7   Wallace (7), 36.
8   Wallace (20), 2:289. I have substituted the modern designations for Wallace's species names.
9   Smith (1), 327.
10   Fichman, 267-68.
11   Wallace (5), 1:401.
12   Wallace (5), 1:401.
13   Wallace (1), 131.
14   Wallace (5), 1:401
15   Wallace (5), 1:402
16   Wallace (14), 716.

[17] Wallace (4), 127-30

[18] Dawkins (4), 32.

[19] Camerini, 150. Wallace's emphasis.

[20] Camerini, 147.

[21] George, 251.

[22] Camerini, 171-72, Wallace's emphasis.

[23] Mallet (4), 1.

[24] Darwin (3), 50-51.

[25] Nicholson, 1:372. Nicholson's emphasis.

[26] Darwin (4), 274.

[27] Edey, 70; Bowler (1), 194.

[28] Mallet (4), 8.

[29] Camerini, 160.

[30] qd. Gould (2), 25.

[31] Hopkins, 80.

[32] Darwin (3), 83-84.

[33] Darwin (3), 83.

[34] Wallace (5), 1:401.

[35] Wallace (4), 429-30.

[36] Eisley (1), 306.

[37] Eccles, 240. Eccles' emphasis.

[38] Wallace (2), 463

[39] Wallace (5), 2:474.

[40] Wallace (2), 474.

[41] Dawkins (4), 19, 36, 199.

[42] Eisley (2), 84-85/

[43] Gould (2), 25.

Chapter 10

[1] Carroll,(1), 27.

[2] Carroll,(2)120.

[3] Harrison, 256-58.

[4] Powell (1), 55.

[5] Claus rejects the interpretation of the Homeric psyche as 'breath of life' and plumps for the Nietzschean and Shavian 'life force'. (Claus 7, 92-97). He sees Homer as purging the word of its possible (as he sees it) original meaning of 'breath', and replacing it with "the 'life-force' category of words (97)."

[6] Havelock (1), 69.

7   Powell (1), 57.
8   Havelock (1), 33, fn. 18.
9   Powell (1), 49.
10 Bely, 93.
11 Lang, 66.
12 Powell (1), 57.
13 Powell, "Homer and Writing" in Morris 4.
14 Stanford, 3, 7.
15 Powell (2), 23–24.
16 Bickel, 49.
17 Cysarz et al.
18 Powell (1), 224.
19 Snell, ix, 20.
20 Snell, 9.
21 Bremmer, 22.
22 Burnet, 245.
23 Onians, 119.
24 Onians, 120.
25 Schlam, 7.
26 Schlam, 7.

BIBLIOGRAPHY

Albert, David Z., Quantum Mechanics and Experience (Cambridge, Mass.: Harvard University Press, 1992).

Altenmüller, Eckhart O., "Music in your Head," Scientific American. New York: Scientific American Special on Mind (2004), 24-31.

Apuleius, Metamorphoses, ed. J. Arthur Hanson (Cambridge, Mass.: Harvard University Press, 1989).

Atwell, John E., Schopenhauer on the Character of the World: the Metaphysics of Will (Berkeley & Los Angeles: University of California Press, 1995).

Ballantine, Christopher, Twentieth Century Symphony (London: Dobson, 1983).

Barrow, John D. (1). The Artful Universe (Oxford: Clarendon, 1995).

_____ (2). The Constants of Nature: from alpha to omega (London: Cape, 2002).

_____ (3). and Frank J. Tipler. The Anthropic Cosmological Principle (Oxford: Clarendon, 1986).

Bates, Henry Walter, Naturalist on the River Amazons (Lodnon: Dent, 1969).

_____, "Contribution to an insect fauna of the Amazon Valley. Lepidoptera: Heliconiidae," Transactions of the Linnean Society of London 23 (1862): 495-566.

Belayeva, Natalya V., History of Insects (Dordrecht: Kluwer, 2002).

Bely, A., Selected Essays of Bely, ed. Steven Cassidy (Berkeley & Los Angeles: University of California, 1985).

Berlioz, Hector, A Critical Study of Beethoven's Nine Symphonies (London: Reeves, 1958).

Bevan, David, (1) André Malraux: towards the expression of transcendence (Kingston: McGill & Queen's University Press, 1986).

_____,(2)

"The Archaic Smile, or Butterflies and Monsters, Mélanges Malraux, 12, 2 (1980), 19-29.

Bickel, Ernst, "Homerischer Seelenglaube," Schriften der Königsberger Gelehrten Gesellschaft, 1925.

Blackmore, Susan, The Meme Machine (Oxford: Oxford University Press, 2000).

Blend, Charles D., André Malraux: tragic humanist (Columbus: Ohio State University Press, 1963).

Bloch, Ernst, Essays on the Philosophy of Music (Cambridge: Cambridge University Press, 1985).

Boardman, Philip, The World of Patrick Geddes (London: Routledge & Kegan Paul, 1978).

Boberg, I.M., "The Tale of Cupid and Psyche," Classica et Mediaevalia 1 (1938): 177-216.

Böhme, Joachim, Die Seele und das Ich in homerischen Epos. Mit eienem Anhang: Vergleich mit dem Glauben der Primitiven (Leipzig and Berline: B.G. Teubner, 1929).

Boisacq, Emile, Dictionnaire Etymologique de la Langue Grecque (Heidelberg: Winter, 1950).

Bowler, Peter J., (1), Charles Darwin: the years of controversy: the Origin of Species and its critics 1859-1882 (London, 1972).

_____(2), The Non-Darwinian Revolution: Reinterpreting a Historical Myth (Baltimore: Johns Hopkins University Press, 1988).

Bremmer, Jan, The Early Greek Concept of the Soul (Princeton: Princeton University Press, 1983).

Brown, Norman O., (1) Apocalypse and/or Metamorphosis (Berkeley & Los Angeles: University of California Press, 1991).

_____, (2) Closing Time (New York: Vintage, 1973).

_____, (3) Hermes the Thief: the evolution of a myth (New York: Vintage, 1969).

_____, (4) Life against Death: the psychoanalytical meaning of history (London: Routledge, Kegan & Paul, 1959).

_____, Love's Body (New York: Vintage, 1966).

Budick, Sanford, Kant and Milton (Cambridge, Mass.: Harvard UP, 2010).

Burkhardt, Frederick and Sydney Smith, eds., The Correspondence of Charles Darwin 5: 1851-55 (Cambridge: Cambridge University Press, 1989).

_____, The Correspondence of Charles Darwin 6: 1856-57 (Cambridge: Cambridge University Press, 1990).

Burnet, J., "The Socratic Doctrine of the Soul," Proceedings of the British Academy (1916), 235-59.

Butler, Samuel, Evolution, Old and New; or the theories of Buffon, Dr. Erasmus Darwin, and Lamarck, as compared with that of Mr. Charles Darwin (London: Hardwicke & Bogue, 1879).

Butterworth, Brian, The Mathematical Brain (London: Macmillan, 1999).

Calvin, William H. and Derek Bickerton, Reconciling Darwin and Chomsky with the Human Brain (Cambridge, Mass.: MIT Press, 2000).

Camerini, Jane R., ed., The Alfred Russel Wallace Reader (Baltimore: Johns Hopkins, 2002).

Carduner, Jean, La Création Romanesque chez Malraux (Paris: Librairie A-G. Nizet, 1969).

Carneiro, Robert L., The Evolution of the Human Mind: from supernaturalism to naturalism—an anthropological perspective (New York: Eliot Werner, 2010).

Carroll, Sean B., (1) "Evo-Devo and an Expanding Evolutionary Synthesis: A Genetic Theory of Morphological Evolution," Cell 134 (July 11, 2008), 25-36.

_____, (2) Grenier, Jennifer K., Weatherbee, Scott D. (eds.), From DNA to Diversity: molecular genetics and the evolution of animal design (Malden, Mass.: Blackwell, 2005).

Cavalli-Sforza, Luigi and Marcus Feldman, Cultural Transmission and Evolution: a quantitive approach (Princeton: Princeton University Press, 1981).

Chambers, Robert, Vestiges of the Natural History of Creation. (Edinburgh: Churchill, 1884).

Chomsky, Noam, (1) New Horizons in the Study of Language and Mind (Cambridge: Cambridge University Press, 2000).

_____, (2), Rules and Representations (Oxford: Blackwell, 1980).

Clack, J., Gaining Control (Bloomington: Indiana University Press, 2002).

Claus, David B., Toward the Soul: an inquiry into the meaning of ψυχή before Plato (New

Haven: Yale University Press, 1981).

Courcel, Marlinel (ed.), Malraux: Life and Work, (London: Weidenfeld & Nicolson, 1977).

Croce, Benedetto, (1) Aesthetic: as science of expression and general linguistic, trans. Douglas Ainslie (London: Vision Press, 1959).

_____, (2). Goethe (London: Read Books, 123).

Cysarz, D.; Von Bonin, D.; Lachner, H.; Moser, M.; Bettermann, H., "Oscillations of heart rate and respiration synchronize during poetry recitation," American Journal of Physiology—Heart & Circulatory Physiology, 287, no. 2 (September 2004): H579.

Darcus, S.M., "A person's relation to ψυχή in Homer, Hesiod, and the Greek Lyric Poets," Glotta 57 (1979): 30-39.

Darwin, Charles (1). The Descent of Man, and selection in relation to sex. (Princeton: Princeton University Press, 1981).

_____(2). The Origin of Species, ed. Gillian Beer, (Oxford: Oxford University Press, 1998)

_____(3), and T.H. Huxley. Autobiographies, (Oxford: Oxford University Press, 1974).

_____(4), and A.R. Wallace. Evolution by Natural Selection, (Cambridge: Cambridge University Press, 1958).

Dawkins, Richard (1), The Greatest Show on Earth: the evidence for evolution (London: Bantam, 2009).

_____(2), The Blind Watchmaker (London: Longman, 1986).

_____(3), The Extended Phenotype: the gene as the unit of selection (Oxford: Freeman, 1982).

_____(4), River out of Eden: a Darwinian view of life (Weidenfeld & Nicolson, 1995).

_____(5), The Selfish Gene (Oxford: Oxford University Press, 1989).

_____(6), "In defense of selfish genes," Philosophy, 56 (1981): 556-73.

Deacon, Terrence, The Symbolic Species: the co-evolution of language and the human brain (New York: Norton, 1997).

Debarbieri, Cesar A., Los Personajes en la Poética de José María Eguren (Lima: Universidad del Pacifico, 1975).

Defries, Amelia, The Interpreter: Geddes-the Man and his Gospel (London: Routledge, 1927).

Dennett, Daniel, C. (1). Consciousness Explained (Harmondsworth: Allen Lane, 1991).

_____(2). Darwin's Dangerous Idea: evolution and the meanings of life (Harmondsworth: Allen Lane, 1995).

_____(3). "Memes and the Exploitation of the Imagination," Journal of Aesthetics and Art Criticism, 48 no. 2 (Spring 1990): 127-35.

Descartes, René, (1), Cogitationes Privatae, ed. Charles Adam and Paul Tannery. In Oeuvres X (Paris: Vrin, 1966): 213-48.

_____ (2), Principles of Philosophy, trans. Miller, Valentine Rodger and Miller, Reese P. (Dordrecht: Reidel, 1983).

_____ (3), Rules for the Direction of the Mind, trans. L.J. Lafleur. (Indianapolis: Bobbs-Merrill, 1961).

Dicke, R.H., "Dirac's Cosmology and Mach's Principle," Nature 192 (November 4, 1961): 440-41.

Dirac, Paul A.M., Directions in Physics (New York: Wiley, 1978).

_____, "A new basis for cosmology," Proceedings of the Royal Society of London A, 165 (1938): 199-208.

_____, "Reply to R.H. Dicke, Nature 192 (November 4, 1961): 441.

Donehower, Bruce, ed., The Birth of Novalis: Frederick Hardenberg's Journal of 1797 with selected letters and documents (New York: State University of New York, 2007).

Edey, Maitland A. and Donald C. Johnson, Blueprints: solving the mystery of evolution (Oxford: Oxford University Press, 1990).

Eccles, John C., Evolution of the Brain: creation of the self (London: Routledge, 1991).

Edey, Maitland A. and Donald C. Johanson, Blueprints: solving the mystery of evolution (Oxford: Oxford University Press, 1990).

Edwards, Mark W., Sound, Sense and Rhythm: listening to Greek and Latin Poetry (Princeton: Princeton University Press, 2002).

Eguren, José María, Obras Completas, ed., Ricardo Silva-Santisteban (Lima: Mosca Azul, 1974).

Einstein, Albert, Relativity: the Special and General Theory (New York: Crown, 1961).

Eisley, Loren (1), Darwin's Century (London: Gollancz, 1959).

_____(2), The Immense Journey (New York: Vintage, 1958).

_____(3), "Alfred Russel Wallace," Scientific American 200:2 (February 1959): 70-83.

Enz, Charles P., No Time to be Brief: A Scientific Biography of Pauli (Oxford: Oxford University Press, 2002).

Evreinoff, Nicolas, The Theatre in Life, trans. Alexander I Nazaroff (London: Harrap, 1927).

Farmelo, Graham, "Physics+Dirac = Poetry," Guardian (February 21, 2002).

Faye, Emmanuel, Heidegger: the introduction of Nazism into Philosophy (New Haven: Yale University Press, 2009).

Ferrera, Lawrence, "Schopenhauer on music as the embodiment of Will," in Jacquette: 183-99.

Fichman, Martin, An Elusive Victorian: the evolution of Alfred Russel Wallace (Chicago: University of Chicago Press, 2004).

Fitch, W. Tecumseh, The Evolution of Language (Cambridge: Cambridge University Press, 2010).

Foucault, Michel, Reader, ed. Paul Rabinow (London: Penguin, 1991).

Franklin, Peter, The Idea of Music: Schoenberg and others (London: Macmillan, 1985).

Franks, Daniel W. and Jason Noble, "Batesian mimics influence mimicry ring evolution," Proceedings of the Royal Society of London B 271 (2004): 191-96.

Frohock, André Malraux and the Tragic Imagination (Stanford: Stanford University Press, 1952).

Gardner, Mary S., The Biology of Invertebrates (New York: McGraw-Hill, 1972).

Gaston, Kevin J., Biodiversity: a biology of numbers and difference (Oxford: Blackwell, 1996).

Gaukroger, Stephen, Descartes: an intellectual biography (Oxford: Clarendon, 1995).

George, Wilma, Biologist Philosopher: a study of the life and writings of Alfred Russel Wallace (London: Abelard-Schuman, 1964).

Gilbert, Lawrence I., Jamshed R. Tata, Burr G. Atkinson, Metamorphosis: postembryonic reprogramming of gene expression in amphibian and insect cells (San Diego, Ca.: Academic Press, 1996).

Girard, René, "L'Homme et le Cosmos dans L'Espoir et Les Noyers de l'Altenburg," PMLA 68 (1953), 49-55.

Goehr, Lydia, "Schopenhauer and the musicians: an inquiry into the sounds of silence and the limits of philosophizing about music," in Jaquette: 200-28.

Goethe, Johann Wolfgang von, (1), Botany (Waltham, Mass.: Chronica Botanica, 1946).

_____, (2), My Life: Poetry and Truth, ed. Thomas P. Saine and Jeffrey L. Sammons (New Jersey: Princeton University Press, 1987).

Gold, Barri J. "The Consolation of Physics: Tennyson's thermodynamic solution," PMLA, 117 no. 3 (May 2002): 449-464. There is much more on energy physics and culture in her book Thermopoetics: energy in Victorian literature and science (Cambridge, Mass.: MIT Press, 2012).

Gould, Stephen Jay, (1) Bully for Brontosaurus: further reflections in natural history (Harmondsworth: Penguin, 1992).

_____, (2) Ever Since Darwin: reflections in natural history (Harmondsworth: Penguin, 1978).

_____, (3) I Have Landed: splashes and reflections in natural history (London: Cape, 2002).

_____, (4) The Panda's Thumb: new reflections in natural history (Harmondsworth: Penguin, 1990).

_____, (5) The richness of Life (London: Vantage, 2007).

_____, (6) "Darwinian Fundamentalism," New York Review of Books 44, no. 10 (June 12, 1997), 34-37.

Graz, Louis, "L'Iliade et la personne," Esprit 28 (1960): 1390-1403.

Grene, Marjorie, Descartes (Indianapolis: Hackett, 1998).

Griffiths, Anthony J.F., Wessler, Susan R., Lewontin, Richard C., Carroll, Sean B. (eds.),

Introduction to Genetic Analysis (New York: Freeman & Co., 2008).

Harris, Geoffrey T., (1), André Malraux: a reassessment (London: Macmillan, 1996).

_____, (2), "Note on Le Miroir des Limbes: 'A Convoy of Utopias and Aspirations,'" Mélanges Malraux, 13, no. 2 (1981), 22-25.

Harrison, S.J., Apuleius: a Latin Sophist (Oxford: Oxford University Press, 2000).

Havelock, Eric A., (1), The Literate Revolution in Greece and its cultural consequences (Princeton: Princeton University Press, 1982).

_____, (2), The Muse Learns to Write: reflections on orality and literacy from antiquity to the present (New Haven: Yale University Press, 1986).

Hegel, G.W.F., The Philosophy of Fine Art, vol 3, trans. F.P.B. Osmaston (London: Bell, 1920).

Heidegger, Martin, On the Way to Language (Unterwegs zur Sprache), trans. Peter D. Hertz (New York: Harper & Row, 1971).

Heller, Erich, Goethe and the idea of scientific truth (Swansea: University College, 1949).

Hewitt, James R., André Malraux (New York: F. Unger, 1978).

Hiebel, Frederick, Novalis: German Poet—European Thinker—Christian Mystic (Brooklyn: AMS Press, 1969).

Higgins, James, The Poet in Peru: Alienation and the Quest for a Super-Reality (Liverpool: Liverpool University Press, 1982).

Holub, Miroslav, The Dimension of the Present Moment and other essays (London: Faber, 1990).

Hopkins, Gerard Manley, Poems, ed. W.H. Gardner and N.H. MacKenzie (Oxford: Oxford University Press, 1970).

Horvath, Violet, André Malraux: the human adventure (New York: New York University Press, 1969).

Husserl, Edmund, (1), Briefwechsel (Dordrecht: Kluwer, 1994).

_____, (2) Cartesian Meditations

_____, (3), The Crisis of European Sciences and Transcendental Phenomenology: an introduction to phenomenological philosophy (Evanston: Northwestern University Press, 1970).

Immisch, Otto, "Sprachliches zum Seelenschmetterling," Glotta 6 (1915): 193-205.

Jacquette, Dale, ed. Schopenhauer, philosophy, and the arts (Cambridge: Cambridge University Press, 1996).

Jones, Steve, (1), Almost Like a Whale: The Origin of Species Updated (London: Black Swan, 2001).

_____, (2), Y: the descent of men (London: Little, Brown, 2002).

_____, (3), Darwin's Island: the Galapagos in the Garden of England (London: Little, Brown, 2009).

Joron, Mathleu and James L.B. Mallet, "Diversity in mimicry: paradox or paradigm?" TREE, 13 no. 11 (November 1998): 462-66.

Kant, I., Critique of Pure Reason (London: Dent, 1969).

Keats, John. The Letters of John Keats 1814-1821 vol. 2, ed., Hyder Edward Rollins (Cambridge: Cambridge University Press, 1958.

Keller, Pierre, Husserl and Heidegger on human experience (Cambridge: Cambridge University Press, 1999).

Khan, Charles H., The Art and Thought of Heraclitus: an edition of the fragments with translation and commentary (Cambridge, England: Cambridge UP, 1979).

Kimura, Motoo, The Neutral Theory of Molecular Evolution (Cambridge: Cambridge University Press, 1983).

Kirk, G.S., "The Structure of the Homeric Hexameter," Yale Classical Studies, 20 (1966), 76-104.

Kline, Thomas Jefferson, André Malraux and the Metamorphosis of Death (New York: Columbia University Press, 1973).

Knox, Israel, The Aesthetic Theories of Kant, Hegel, and Schopenhauer (New York: the Humanities Press, 1958).

Kohn, David, ed. The Darwinian Heritage (Princeton: Princeton University Press, 1985).

Kottler, Malcolm Jay, "Charles Darwin and Alfred Russel Wallace: two decades of debate over Natural Selection," in Kohn, 367- 432.

_____, "Alfred Russel Wallace," Isis, 65 (1974), 144-92.

Kuhn, Thomas, The Structure of Scientific Revolutions (Chicago: University of Chicago Press, 1996).

Kurzweil, Ray, The Singularity is Near: when humans transcend biology (London: Duckworth, 2009).

Larson, Edward J., Evolutions's Workshop: God and Science on the Galápagos Islands (Harmondsworth: Allen Lane, 2001).

Layton, Robert, ed. A Guide to the Symphony (Oxford: Oxford University Press, 1995).

Leibniz, G.W., "Principles of Nature and of Grace, based on Reason," Philosophical Papers & Letters 2, ed. Leroy E. Loemker, (Dordrecht: Reidel, 1976).

Leonard, Miriam, (ed.), Derrida and Antiquity, (Oxford: Oxford UP, 2010).

Le Sage, Laurent, The New French Novel: an interpretation and sampler (State College: Pennsylvania State University, 1962).

Lesser, Friedrich Christian, Théologie des Insectes, ou Demonstrations des Perfections de Dieu (La Haye: Swart, 1742).

Levin, Simon Asher ed., Encyclopedia of Biodiversity 5 (San Diego: Academic Press, 2000).

Levine, Suzanne Jill, "Afterwords on Afterthoughts," Review of Contemporary Fiction, 10 no. 3 (Fall 1990): 181-82.

Levy, Karen Dydo, "André Malraux and the Farfelu: Quest for transcendence," Melanges Malraux, 4 no. 1 (Spring 1972), 30.

Lindorff, David, Pauli and Jung: the meeting of two great minds (Wheaton, Ill: Quest Books, 2004).

MacDiarmid, Hugh, In Memoriam James Joyce; A Vision of World Language (Glasgow: MacLellan, 1955).

Magee, Bryan (1), Misunderstanding Schopenhauer (London: Institute of German Studies, 1990).

_____ (2), The Philosophy of Schopenhauer (Oxford: Clarendon, 1983).

Maldacena, Juan, "Into the fifth dimension," Nature 423 (12 June 2003): 695-96.

Mallet, James L.B., (1), "Wallace and the Species Concept of the Early Darwinians," in Smith (2): 102–113.

_____(2), "The genetics of biological diversity: from varieties to species," Gaston: 13–53.

_____(3), "Mimicry meets mitochondrion," Current Biology, 6, no. 8: 937-40.

_____(4), "Poulton, Wallace and Jordan: how discoveries in Papilio butterflies initiated a new species concept 100 years ago," Systematics and Biodiversity (2004): prepublication, 1-14.

_____(5), "Concepts of Species" in Levin: 427-40.

Malraux, André, Anti-Memoirs, (1), trans. Terence Kilmartin (New York: Holt, Reinhart & Winston, 1968).

_____, La Condition Humaine (2), (Paris: Gallimard, 1933).

_____, L'Espoir (3) (Paris: Gallimard, 1937).

_____, L'Homme précaire et la Litérature (4), (Paris: Gallimard, 1977).

_____, Fallen Oaks: conversations with De Gaulle (5), (New York: Holt, Reinhart & Winston, 1971).

_____, D'une Jeunesse Européenne (6), (Paris: Grasset. 1927).

_____, Lunes en papier (7), (Paris: Simon, 1921).

_____, Le miroir des limbes (8), (Paris: Gallimard, 1975). and Volume 2, La corde et le souris (Paris: Gallimard, 1976).

_____, Les Noyers de \l'Altenburg (9), (Paris: Gallimard, 1948).

_____, La Voie Royale (10), (Paris: Grasset, 1930).

_____, Les Voix du Silence (11), (Paris: Gallimard, 1951). The Voices of Silence, trans. Stuart Gilbert (New York: Doubleday, 1953).

_____, The Temptation of the West (12), (Chicago: Chicago University Press, 1992).

_____, "Jeune Chine," (13) NRF, 38 (January 1932), 5-7.

Malraux, Clara, Nos Vingt Ans (Paris: Grasset, 1960).

Margulis, Lynn, The Symbiotic Planet: a new look at evolution (New York: Basic Books, 1998).

_____ and Dorion Sagan, Acquiring Genomes: a theory of the origin of species (New York: Basic Books, 2002).

Mayr, Ernst, (1) Evolution & the Diversity of Life: selected essays (Cambridge, Mass: Belknap, Harvard UP, 1976).

_____(2) and Provine, W. (eds.), The Evolutionary Synthesis (Cambridge, Mass: Harvard UP, 1980).

McKinney, H. Lewis, Wallace and Natural Selection (New Haven: Yale University Press, 1972).

Meier, C.A., ed. Atom and Archetype: the Pauli-Jung Letters, 1932-1958 (Hove: Routledge, 2011).

Mellers, Wilfrid, The Sonata Principle (London: Barrie & Jenkins, 1973).

Menn, Stephen, Descartes and Augustine (Cambridge: Cambridge University Press, 1998).

Miller, Arthur I., Deciphering the Cosmic Number: the strange friendship of Wolfgang Pauli and Carl Jung (New York: Norton, 2009).

Molnár, Géza, Novalis 'Fichte Studies:' the foundation of his aesthetics (The Hague: Mouton), 1970).

Müller, Max, Lectures on the Science of Language (New York: Scribner, 1862).

Nabokov, Vladimir, Invitation to a Beheading (London: Penguin, 1963).

_____, The Stories (New York: Vintage, 1997).

Nancy, Jean-Luc, The Birth to Presence (Stanford: Stanford University Press, 1993).

Navaud, Guillaume, Persona: le théâtre comme métaphore théorique de Socrate à Shakespeare

(Genève: Droz, 2011).

Nicholson, A.J., Evolution after Darwin, ed. S. Tax, (Chicago: University of Chicago Press, 1960

Novalis, (1) The Disciples at Saïs and other fragments (London: Methuen, 1903).

_____, (2) "Pollen (Blüthenstaub)," New Literary History 22 (1996).

Nuñez, Eduardo, José María Eguren: Vida y Obra (Lima: n.pub.1964).

O'Brien, William Arctander, Novalis: signs of revolution (Durham: Duke University Press, 1995).

O'Connell, Erin, "Derrida and Presocratic Philosophy," in Leonard.

O'Meara, Dominic J., Pythagoras Revived: mathematics and philosophy in late antiquity (Oxford: Clarendon, 1989).

Onians, Richard Broxton, The Origin of European Thought: about the body, the soul, the world, time, and fate (Cambridge: Cambridge University Press, 1951).

Ortega, Julio, ed., Eguren: Antología (Lima: Editorial Universitaria, n.d.).

Palmquist, Stephen R., Kant's System of Perspectives: an architectonic interpretation of the critical philosophy (Lanham, Md.: University Press of America, 1993).

Pauli, Wolfgang, Writings on Physics and Philosophy: ed. Charles P. Enz and Karl von Meyenn, trans. Robert Schlapp (Berlin: Springer-Verlag, 1994).

Paz, Octavio, Claude Levi-Strauss o el Nuevo Festin de Esopo (Mexico: Joaquín Mortiz, 1969).

Penrose, Roger, The Road to Reality: a complete guide to the laws of the universe (London: Cape, 2004).

Pfefferkorn, Kristin, Novalis: a romantic's theory of language and poetry (New Haven: Yale University Press, 1988).

Pinker, Steven, How the Mind Works (London: Allen Lane, 1997).

Powell, Barry B. (1). Homer and the origins of the Greek Alphabet (Cambridge: Cambridge University Press, 1991)

_____ (2). Writing and the origins of Greek Literature (Cambridge: Cambridge University Press, 2002).

Prudhomme, Benjamin and Nicolas Gompel, "Genomic hourglass," Nature (2010).

Punnett, R.C., Mimicry in Butterflies (Cambridge: Cambridge University Press, 1915).

Quiring, R., Walldorf, U., Kloter, U., Gehring, W.J.: "Homology of the eyeless gene of Drosophila to the Small eye gene in mice and Aniridia in humans." Science (1994), 785-9.

Raby, Peter, Alfred Russel Wallace: A Life (London: Chatto & Windus, 2001).

Radulescu, Domnica, André Malraux: the 'farfelu' as expression of the feminine and the erotic (New York: Peter Lang, 1994).

Rasnitsyn, Alexander P. and Donald L.J. Quicke, eds., History of Insects (Dordrecht: Kluwer, 2002).

Rees, G.O., "Animal Imagery in the Novels of André Malraux," French Studies, 9: no. 2 (April 1955), 129-42.

Rees, Martin, Just Six Numbers: the deep forces that shape the universe (London: Weidenfeld & Nicolson, 1999).

Ríos, Julián. (1) Larva: Midsummer Night's Babel, trans. Richard Alan Francis (London: Quartet, 1991).

_____, (2) Review of Contemporary Fiction (1990), 10:3, 182.

Rohde, Erwin, Psyche: the cult of souls and belief in immortality among the Greeks (London: Kegan Paul, Trench, Trubner & Co., 1925).

Rosen, Charles, The Classical Style: Haydn, Mozart, Beethoven (London: Faber, 1976).

Russell, Bertrand, Philosophical Essays (London: Longman Green, 1910).

Ruxton, G.D., M. Speed and T.N. Sherratt, "Evasive Mimicry: when (if ever) could mimicry based on difficulty of capture evolve?" Proceedings of the Royal Society of London B 271 (2004): 2135-42.

Saul, Nicholas, ed., Companion to German Romanticism (Cambridge: Cambridge University Press, 2009).

Schelling, F.W., The Philosophy of Art, ed. and trans. Douglas W. Scott. Foreword by David Simpson (Minneapolis: University of Minnesota Press, 1989).

Schlam, Carl, Cupid and Psyche: Apuleius and the Monuments (Pennsylvania: American Philological Association, 1976).

Schlipp, Paul Arthur, ed. Albert Einstein: Philosopher-Scientist (Lasalle, Ill.: Open Court, 1949).

Schopenhauer, Arthur (1). Manuscript Remains in 4 volumes, ed. Arthur Hübscher. trans. E.F.J. Payne. (Oxford: Berg, 1988-90).

_____(2). On the Will in Nature: A Discussion of the Corroborations from the Empirical Sciences that the Author's Philosophy has received since its first appearance. trans. E.F.J. Payne. edited with a preface by David E. Cartwright. (Oxford: Berg, 1992).

_____(3). Parerga and Paralipomena: short philosophical essays. trans. E.F.J. Payne (Oxford: Clarendon, 2000).

_____(4). Schopenhauer's early Fourfold Root. trans. and commentary F.C. White (Aldershot: Avebury, 1997).

_____(5). The World as Will and Representation. trans. E.F.J. Payne. (New York: Dover, 1958).

Shubin, Neil, Your Inner Fish: a journey into the 3.5 billion-year history of the human body (New York: Pantheon, 2008).

_____, C. Tabin, and S. Carroll, "Fossils, genes, and the evolution of animal limbs," Nature 388 (1997): 639-388.

Simpson, David, The Origins of Modern Critical Thought: German aesthetic and literary criticism from Lessing to Hegel (Cambridge: Cambridge University Press, 1988).

Snell, Bruno, The discovery of the Mind in Greek Philosophy and Literature (New York: Dover, 1982).

Spengler, Oswald, The Decline of the West (London: Allen & Unwin, 1926-8).

Smith, Charles H., (ed.), (1) Alfred Russel Wallace: and anthology of his shorter writings (Oxford: Oxford University Press, 1991).

Smith, Charles H. and George Beccaloni, (eds.) (2) Natural Selection and Beyond: the intellectual legacy of Alfred Russel Wallace (Oxford: Oxford University Press, 2008).

Solso, Robert L., The Psychology of Art and the Evolution of the Conscious Brain (Cambridge Mass.: MIT, 2003).

Spiegelberg, Herbert, The Phenomenological Movement—a historical introduction (The Hague: Nijhoff, 1969).

Spooner, David, (1) The Insect-Populated Mind: how insects have influenced the evolution of consciousness (Lanham, Md.: Hamilton Books, 2005).

——————, (2) The Metaphysics of Insect Life and other essays (San Francisco: International Scholars, 1995).

——————, (3)"The Response of Some Writers to the Spanish Civil War," (Ph.D. Bristol University Library).

Stanford, W.B., The Sound of Greek: studies in Greek theory and practice of euphony (Berkeley & Los Angeles: University of California Press, 1967).

Steiner, George (1), After Babel: aspects of language and translation (Oxford: Oxford University Press, 1975).

——————(2), Extraterritorial: papers on literature and the language revolution (Harmondsworth: Penguin, 1975).

——————(3), Grammars of Creation (London: Faber. 2001).

Sunnen, Myriam, Malraux et le Christianisme (Paris: Champion, 2009).

Tanner, Michael, Schopenhauer: Metaphysics and Art (London: Phoenix, 1998).

Tannery, Claude, Malraux: the absolute agnostic, or metamorphosis as universal law (Chicago: University of Chicago Press, 1991).

Tax, S., ed. Evolution after Darwin (Chicago: University of Chicago Press, 1960).

Thomas, J.A. et al., "Comparative Losses of British Butterflies, Birds, and Plants and the Global Extinction Crisis," Science 303 (19 March 2004): 1879-81.

Thoreau, Henry D., A Week on the Concord and Merrimack Rivers, ed. Carl F. Hovde, William L. Howarth and Elizabeth Hall Witherell (Princeton: Princeton University Press, 1980).

Todorov, Tzvetan, Theories of the Symbol, trans Catherine Porter (Oxford: Blackwell, 1982).

Truman, James W. and Lynn M. Riddiford, "The Origins of Insect Metamorphosis." Nature 41 (30 September 1999): 447-52.

Vandegans, André, La Jeunesse Littéraire d'André Malraux: essai sur l'inspiration farfelue (Paris: Jean-Jacques Pauvert, 1964).

Vogel, Stanley M., German Literary Influences on the American Transcendentalists (New Haven: Yale University Press, 1955)

Vorzimmer, Peter J., Charles Darwin: the years of controversy: The Origin of Species and its critics 1859-1882. (Philadelphia: Temple University Press, 1972).

Wallace, Alfred Russel (1). Contributions to the theory of Natural Selection. London: Macmillan, 1871.

_____(2). Darwinism: an exposition of the Theory of Natural Selection with some of its applications. London: Macmillan, 1889.

_____(3). Island Life, or the phenomena and causes of Insular Faunas and Floras: including a revision and attempted solution of the problem of geological climates. London: Macmillan, 1892.

_____(4). The Malay Archipelago: the land of the orang-utan and the bird of paradise. A narrative of travel, with studies of man and nature. 5th ed. London: Macmillan, 1894.

_____(5). My Life: a record of events and opinions. London: Chapman & Hall, 1905.

_____(6). A Narrative of Travels on the Amazon and Rio Negro: with an account of the native tribes, and observations on the climate, geology and natural history of the Amazon Valley. London: Ward, Lock & Co., 1890.

_____(7). Natural Selection and Tropical Nature: Essays on descriptive and theoretical biology. London: Macmillan, 1891.

_____(8). The Wonderful Century: its successes and failures. London: Swan Sonnenschein, 1898.

_____(9). The World of Life: a manifestation of creative power, directive mind and ultimate purpose. London: Chapman & Hall, 1910.

_____(10). "On the Pieridae of the Indian and Australian Regions." Transactions of the Entomological Society of London, 3rd ser., 4, part III: 301-416.

_____(11). "Description of a New Species of Ornithoptera. Ornithoptera brookiana." Proc Entomol Soc Lon (1854-55): 104-105.

_____(12). "Geological Climates and the Origin of Species." Quarterly Review, 126 (1869): 391-394. Also in Wallace (7).

_____(13). "Limits of Natural Selection as applied to man." In Wallace (1).

_____(14). "Mimicry and other Protective Resemblances among Animals." Westminster Review, 32 (n.s.), no. 1: 1-43.

_____(15). "Natural Selection." Athenaeum, no. 2040 (1 December 1866): 716-717.

_____(16). "On the Entomology of the Aru Islands." Zoologist 16 (January 1858), nos. 185-186: 5889-94.

_____(17). "On the Habits of the Butterflies of the Amazon Valley." Trans Entomol Soc Lon 2 (n.s.): pt. 8 (April 1854): 253-64.

_____(18). "On the Law which has regulated the Introduction of New Species." Annals and Magazine of Natural History 16 (2nd ser.) (September 1855): 184-96.

_____(19). "On the Tendency of Varieties to Depart Indefinitely from the Original Type." Journal of the Proceedings of the Linnean Society: Zoology 3 (9) (20 August 1858): 45-62.

_____(20). "The Phenomenon of Variation and Geographical Distribution as Illustrated by the Papilionidae of the Malayan Region." Transactions of the Linnean Society of London, 25, pt. 1 (1865): 1-71.

_____(21). "Protective Mimicry in Animals." in Science for All 2, ed. Robert Brown. London: Cassell, Petter, Galpin & Co., 1879: 284-96.

_____(22). "Regarding Mimicry in Insects." J Proc Entomol Soc Lon (1864): 14-15.

Wenzl, Aloys, "Einstein's Theory of Relativity, viewed from the standpoint of its critical realism, and its significance for philosophy," in Schlipp, 581-606.

Wesson, Robert, Beyond Natural Selection (Cambridge, Mass.: MIT, 1991).

Wheeler, Kathleen M., ed., German aesthetic and literary criticism: the Romantic ironists and Goethe (Cambridge: Cambridge University Press, 1984).

Wigglesworth, Vincent, The Life of Insects (London: Weidenfeld & Nicolson, 1964).

Williams-Ellis, Amabel, Darwin's Moon: a biography of Alfred Russel Wallace (London: Blackie, 1966).

Wittgenstein, Philosophical Investigations, ed. G.E.M Anscombe and R. Rhees (Oxford: Blackwell).

Wood, Rupert, "Language as Will and Representation: Schopenhauer, Austin, and Musicality," Comparative Literature 48 no.4 (Fall 1996): 302-25.

Woodcock, George, W.H. Bates: naturalist of the Amazon (London: Faber, 1969).

Wrathall, Mark and Jeff Malpas (eds.), Heidegger, Authenticity, and Modernity: Essays in Honor of Hubert C. Dreyfus Vol. 1 (Cambridge, Mass.: MIT, 2000).

Yourgrau, Palle, The disappearance of time: Kurt Gödel and the Idealistic Tradition in Philosophy (Cambridge: Cambridge University Press, 1991).